SYSTEMS GRAPHICS

THE McGRAW-HILL DESIGNING WITH SYSTEMS SERIES

Stitt   *Systems Drafting: Creative Reprographics for Architects and Engineers* (1980)

Daryanani   *Building Systems Design with Programmable Calculators* (1980)

Stitt   *Systems Graphics: Breakthroughs in Drawing Production and Project Management for Architects, Designers, and Engineers* (1984)

Teicholz   *CAD/CAM Handbook* (1984)

Orr   *The Architect's CADD Primer* (1984)

Bakey   *CAD in Architecture, Engineering, and Construction* (1984)

Ouye/Stitt   *Architectural Planning and Design with Systems* (1984)

ERIC TEICHOLZ, CONSULTING EDITOR

SYSTEMS GRAPHICS

BREAKTHROUGHS IN DRAWING PRODUCTION AND PROJECT MANAGEMENT FOR ARCHITECTS, DESIGNERS, AND ENGINEERS

FRED A. STITT

McGRAW-HILL BOOK COMPANY

New York   St. Louis   San Francisco   Auckland
Bogotá   Hamburg   Johannesburg   London
Madrid   Mexico   Montreal   New Delhi
Panama   Paris   São Paulo   Singapore
Sydney   Tokyo   Toronto

TO A MAN WHO GAVE EVERY MOMENT OF HIS LIFE TO
HIS BELOVED PROFESSION. A GREAT TEACHER AND
ARCHITECT, BRUCE GOFF. 1904—1982.

Library of Congress Cataloging in Publication Data
Stitt, Fred A.
  Systems graphics.

  (McGraw-Hill designing with systems series)
  Includes index.
  1.Drawing-room practice.   2.Copying processes.
3.Engineering graphics.   I.Title.   II.Series.
T352.S74   1984   604.2   83-866
ISBN 0-07-061551-9

234567890   HAL/HAL   898765

ISBN 0-07-061551-9

The editors for this book were Joan Zseleczky and William
B. O'Neal, the designer was Mark E. Safran, and the
production supervisor was Sally Fliess. It was set in
Melior by Santype-Byrd.

Printed and bound by Halliday Lithograph.

The comparatively few references to women practitioners in
this text in no way suggest bias on the part of author or
publisher but are a reflection of the current situation in A/E
professions.

CONTENTS

CONTENTS

PREFACE

Many hundreds, possibly thousands, of architects and engineers now routinely do much of their design development and working drawings 30 to 40 percent faster than is possible with traditional drafting. They not only do it faster, they do it less expensively, 30 to 40 percent so. That goes for big firms, one-person offices, architects, engineers, interior designers, specialized designers, and drafters in every discipline, both in the United States and abroad.

Many of these people started straight from my first book, *Systems Drafting—Creative Reprographics for Architects and Engineers.* They have used that information as a springboard for extraordinary new levels of experimentation and innovation. You'll see some of their comments in the first chapter.

Most of the major savings come with new, inexpensive forms of in-house reprographics. You'll see these techniques in this book, all clearly illustrated step by step. It is hoped you'll come to use them yourself. And as you do, you'll be surprised to meet people who will tell you that what you are doing can't be done. "You can't make a reproducible of a paste-up sheet for only $5. It costs us $50 to $100." "You can't make a screened shadow print without sending it out." "You can't make transparencies out of opaque originals without using photography." "You can't run paste-ups or appliqués in a diazo machine without getting ghost shadows."

I know you'll hear such things. I hear them all the time, and after I show someone exactly how it all works, they sometimes will look confused and say those things all over again.

"Systems," as ever, means an integration of different elements. People fail with systems or get less than top results when they get a mind lock on some particular aspect of systems and lose touch with the rest. All aspects of systems are great, but mainly so in relationship with the other aspects. When you hear arguments over the relative importance of paste-up versus overlay, or photographic versus contact printing, or reprographics versus computer design, you're hearing expressions of mind lock. The tools are all vital parts of a totality we call "systems graphics." They all work best when used together to augment one another. They all lose value quickly for those who use them on the wrong jobs. They especially lose value in a hurry for those who are slow to respond to change and who fail to integrate newer tools with what they're already using.

The in-house paste-up and reproducible print methods are great on many jobs, but they have their limits. Some projects use little or no paste-up but accrue a major time or cost advantage, or both, by using overlays.

Overlay will save 20 percent or more of production time and costs on some jobs (many consulting engineers do even better than that). And it is an absolute boon in improving quality control and coordination among consultants and construction trades. But it has its limits too. And it's often used in the wrong ways on the wrong projects not because of inherent limits but because of very common human misunderstandings.

The computer can bring you efficiency ratios of 20 to 1 in many special applications. But it's the most badly used of the new tools. It has to be approached with an extremely focused and organized frame of mind, which, if I dare say so, hasn't been all that characteristic of certain of our design professions.

In-house reprographics, astonishingly productive on some jobs, just isn't that good on others.

Often enough, the out-of-house reprographic service will do wonders for you. Sometimes they can't. Photo systems, photocopying systems, paste-up systems, overlay systems, computer systems—they're all great in their place, and all have limits. But when you put them together, you create a powerhouse.

The trick is to combine advanced systems graphics and the computer so that each is used where most effective. Some computer-aided design and drafting (CADD) advocates say the computer is all you need. That may be so for certain limited schematic linear work such as in electrical engineering and piping diagrams. But the advice fails totally in the multidisciplined, multifaceted world of building design and construction. There are lots of different kinds of tasks, and you need lots of different tools to do them properly, tools ranging from pencil and sketch paper to the cathode ray tube (CRT) and electric stylus.

Tools constitute only half of the systems approach. Project management is the other half. When you combine advanced graphics with the computer and then combine them both with advanced management techniques (systems management), you'll experience a level of productivity and control over your work that you may not right now dream possible.

Some while back I was pleased to run into systems pioneer Ned Abrams at one of the famous systems conferences and trade shows. He said, "It's incredible. We've come this far in just five years." It was true; we had come an incredible distance. He and I stood in a cavernous room that showed it all: computer terminals at every turn, high-speed plotters, vacuum frames of all kinds, electronic lettering machines, and a seemingly endless display of word processors, microcomputers, and minicomputers. Ned and I spoke briefly, verbally patting each other on the back for being a part of this revolutionary change. Shortly afterward a young and very earnest-looking young man came up to me and blurted out an introduction about being an architect concerned about human values. Then he said, "Now we have the technology, Mr. Stitt. But what's it going to do for us? Or to us?"

Good question, I thought. I hoped to pacify the rather agitated architect with some platitudes about efficiency and better design and better quality control. He knew, as I did, that those were canned answers, just stock remarks from my lectures. When we ended the conversation, we both felt unsatisfied. He was trying to get at a much larger issue, and I hadn't dealt with it.

The question is: What really is going to happen to us by using these systems, and will it really be to the good? This book has answers, and some of them add up to a whopping big surprise.

FRED A. STITT

SYSTEMS GRAPHICS

PART 1
ADVANCED IN-HOUSE REPROGRAPHICS

Illustration by Lanée La 'shagway.

SYSTEMS DRAFTING AND SYSTEMS GRAPHICS: WHERE ARE WE NOW?

SUCCESS STORIES

"I feel like we've been drawing on stone slabs with berry juice." That was one architect's response to a recent systems drafting workshop. It's an understandable response. Some firms have advanced so much in just the past couple of years that traditional offices seem a generation behind by comparison. Even newcomers to systems technology report startling results as they try new methods. Here are some examples described by participants in my lectures and at the University of Wisconsin extension's annual conferences on drawing production:

- A Phoenix architectural and engineering firm budgeted 2400 drafting hours on a county government building project and finished in 1300 hours. That was its first all-out systems job—using keynoting, typewritten notes, and composite and overlay drafting.

- A Connecticut architect scheduled 1600 hours to produce a nursing home, but wrapped it up in 900 hours. The systems approach used: mainly paste-up and overlay, using an in-house vacuum frame.

- A two-person firm in Ohio estimated three months production time on a housing project and finished in less than six weeks. The firm used a systems approach centering on paste-up, using a vacuum frame. Concerning a similar project that followed, the architect said, "It was so easy, I almost felt embarrassed to accept the fee."

What's the magic behind these spectacular improvements? It's all wrapped up in one concept: reusability. Most of what we create is reusable in one way or another. And most of what we do is, in one way or another, a repeat of what we've done before.

Take drafting as an example. *All* drafting is repeat drafting. *All* drafting is the recreation of information that already exists. Every note, every title, every north arrow, every door swing, every line has been drawn before and will be drawn again by thousands of designers and drafters. The problem has been one of how to store and retrieve all that original data if it is reusable, instead of recreating it from scratch.

Now we have solutions to that problem. Effective solutions. Low-cost solutions. Much more so even than when I first wrote the book *Systems Drafting*.

THE NECESSITY FOR CHANGE

Let's look at the situation on another level. Look in on a typical drafting room and you're likely to see the following kinds of events:

1. A building designer has just gotten the go-ahead to proceed from the client and pass a project on to working drawings. So the designer hands over design development drawings showing the site plan, floor plans, elevations, and a couple of cross sections. The production team will now take those plans, elevations, etc., and, no matter how accurate the originals are, redraw them from scratch. Why? Because the originals are covered top to bottom with rendering entourage: furniture, oversized titles, shades, shadows, textures, and all the other images used for presentation drawings. Those drawings could have been reused directly in working drawings, but they weren't made with that in mind, and so they can't be. They have to be redrawn.

In contrast, many offices now routinely begin their working drawings as a continuation of design development drawings. They don't start over, and, thus, when they begin production, they're already automatically 20 percent finished. That's a big head start that can mean the difference between success and failure on a job.

2. During the development of detail drawings and schedules, it becomes obvious that the sheets haven't been well planned. Information has to be moved, a detail moved over, a schedule inserted, other details erased.

Instead of picking up a detail drawing that has to be moved and moving it, the drafter has to make a print, erase the original, and then trace and redraw the detail in its new location. The same is true with the schedules to be moved. They were redrawn from other samples to begin with. Now they're erased. Then they're redrawn in new places on the drawing sheets. It's absolute madness.

3. And it goes on. A designer is a little rushed and doesn't check the latest zoning restrictions, so the building drawn on the site plan has to be moved over 5 feet. Instead of picking up the building and moving it, which would be the sensible way of making the change, the designer erases it and redraws it in the new position. And

later, because of the need to widen a service driveway, it will be erased and redrawn again.

4. The redrawing disease is spread onward from architect to engineer. The architects may have created perfectly good plans and sections for the engineers. But the engineers' drafting staffs receive prints which they have to copy. They redraw the building from scratch and, likely as not, misdraw it in the process.

Those were the kinds of things that confronted me when I first entered the drafting room 20 years ago. And they're still more common than not. The nation is changing to systems, but most design firms subject their staffs to exactly that same level of plodding, tedious, erase-and-redraw process. And if anyone wonders why design professionals, architects in particular, are so poorly paid, you can begin to glimpse the answer.

It was this kind of experience and observation that drove me into research and education. It was clear that the profession I loved was in trouble and dearly needed some help.

IT ALL ADDS UP

Bosses of architecture and engineering firms have argued: "Sure you can save a few minutes here or there, but it's a whole lot of trouble to do it. And what difference does it make anyway?"

I had that very statement thrown at me a while back by a semiretired architect-engineer. He runs a part-time practice in Berkeley—pretty much a one-man show, as it always has been throughout his working life.

I was visiting and poking through his files at random while he was working on some wall sections. I noticed that the sections he was doing now were identical to others in previous drawings. So I showed these drawings to him and said, "You've been drawing the same wall sections for years. Why don't you set up a standards system so you can just reuse them directly?"

No one appreciates unsolicited advice, and he was no exception. He exploded: "Just can it, Stitt. I know all about your bloody systems. I know I could save a few minutes on these jobs, but what the hell difference does it make?"

I didn't quite know how to answer at the moment because of the magnitude of the real answer. The truth is that those who go to the fuss required to start "saving a few minutes" start to salvage 5 percent, then 10 percent, then up to 30 and 40 percent of time and cost over traditional drafting. I don't make up these numbers; it's the constant story I get everywhere from people who have gone the whole route. It shows there's always been that much fat in the traditional pencil-pushing process.

So "what difference" a few minutes makes in this case is that we have a fellow who has been pushing a pencil for 30 years. Ten years of that time could have been spent doing a third more work to augment income, or traveling, or pursuing self-improvement, leisure, or what have you. Ten years! That's the difference "a few minutes" can make over the years.

And, in truth, a large part of the impressive long-term time and money savings of systematic procedures is bound up in what may seem to be trivial little time-savers. Shaving a few seconds here and there on a drafting operation or design drawing won't seem like much. But if that operation is repeated many times, day in and day out, by a drafting room full of people, the total time involved is significant after all.

Here are some more specific cases, all publicly stated by architects and engineers at various systems workshops and lectures:

- A housing architect tells of completing 54 drawings for a 60-unit project in 3½ weeks with two staff members. Another reports doing permit drawings for 1800 units—five towers—in 10 days with seven people. Reuse of design drawings, translucent paste-up techniques, overlay drawings, and in-house vacuum frame contact printers are credited for the extraordinary work flow.

- It took an average of 10.8 work hours per drawing for a custom condominium project, according to Ronald Fash of Rapp Fash Sundin, Inc., of Galveston, Texas. The 24-inch by 36-inch drawings would take from 30 to 40 hours using traditional methods, according to Mr. Fash.

- Slightly over 17 hours per drawing were required for a recent $6.5 million multischool project, according to Michael Goodwin of Tempe, Arizona. Architectural drawings were completed by one project architect and a drafting assistant. The main time-savers included intensive production planning, typewriting and keynoting of all notation, plus appliqué drafting and photo composite paste-up and overlay drafting.

- Forty-five architectural drawings for a medical facility were finished in 900 work hours—less than 20 hours per sheet—at the office of Ronald T. Aday, Inc., in Pasadena, California. The firm estimated that normal drafting time without using systems would have been typically 35 to 40 hours per sheet.

- "Around 38 percent savings" was the estimate by Ed Powers of Gresham, Smith and Partners in Nashville. Besides extensive use of composite and overlay drafting, they've created a comprehensive detail file. With the file they can produce composite detail sheets in about 20 hours for institutional buildings for which detail sheets used to require 100 hours.

- A 1978 General Services Administration cost comparison study suggests even greater savings. A value management study of alternative methods of producing drawings for a $2.5 million alterations and restoration job concluded that traditional drafting and materials would cost $34,320, paste-up drafting would cost $15,190, and overlay drafting would cost $13,226. (The third figure includes $2754 in materials and repro costs as opposed to $120 in materials and printing costs for the traditional method.)

Some firms tell us they've made no major reductions in project time. But they do much more work within traditional time frames—more design, more refinements, more detail studies, etc. Many firms now achieve both—more work and faster delivery. The owner of one medium-sized firm, using paste-up and overlay techniques, can now produce four complete 24-inch by 30-inch presentation drawings in three hours. He says the graphic techniques, plus tight financial management, have more than tripled their annual profit margin.

These are all signs of a new production revolution. And it's not just in drafting. There have been major changes lately in every aspect of the design professions. It's a whole new game in marketing, financial management, quality control, corporate organization, and real estate development, to name a few.

New techniques require new data. That means education. Education for designers, drafters, and managers at all levels. So architects and engineers are heading "back to school" in droves. Their "schools" aren't like the old ones: they include resort retreats, hotels, and drafting "laboratories." The classes range from high-priced professional seminars to informal, brown bag, lunchtime "professional development" courses for office staff and management. Many architectural and engineering firms hire professional educators and turn their studios into classrooms for daylong training sessions and "motivators" on everything from project management to creative problem solving.

Teaching materials are as varied as the topics: packaged slide shows, tape cassettes for both office use and "learning while driving" correspondence courses. And coming up of course: video tapes and disks and computerized programmed teaching. More on how all this works will be presented in the next chapter, Educating Yourself and Staff Members in Systems.

Why bother with Systems is an oft pondered question by Architects and Engineers. Many professionals view the implementation of Systems as too bothersome to allot time for training personnel in the use of new techniques or they feel that changing methods would disrupt office production and morale. Others view Systems as being too expensive, requiring a large outlay of dollars for new equipment and materials.

There is an element of fact in all the above reasons for not implementing Systems. Introducing Systems and training of personnel does require time, can be disruptive to present office procedures and will require an expenditure of dollars for new materials and equipment. However, the transition can be smooth and profitable to the firm that makes a full commitment to changing present methods and carefully plans each step for implementing the system.

Fully 75% of the A/E firms nationwide and approximately 40% of those in Canada have had some exposure through seminars, workshops or books and manuals on the subject of overlay drafting. Other firms have experienced exposure to techniques such as cut and paste and photo-drafting. But according to a survey conducted by the Photo Products Division of E. I. DuPont de Nemours Corporation of Wilmington, Delaware, only about 12% of all A/E firms are actually using Systems officewide; others are using it only on select projects.

Systems is paying off for offices which are using it extensively. Professionals have found that costs and errors have been reduced and many have discovered an added bonus—Systems has provided them with a new marketing tool. . . . Systems is being used by progressive firms as a means of securing new clients.

Another bonus for professionals who are applying Systems has been that the transition to computer-aided drafting and design was simplified. Firms which have not used Systems Drafting techniques such as overlay, cut and paste, photo-drafting and keynoting with standard details, word processing and other automated systems have found the changeover to interactive graphics time-consuming and, in some cases, a failure. The discipline that is required for personnel to plan and create drawings using Systems Drafting prepares a firm for CADD.

Architects and Engineers who have not progressed to CADD but who are applying Systems know that the longer a job is on the boards the smaller the profit. Working drawings must be produced in the shortest time possible to insure a profitable commission. The time saved in the construction documents phase can be given to the design phase. Systems Drafting shortens the time

required to produce working drawings to about 30% of the total project budget. Gresham, Smith and Partners has revised the Standard AIA breakdown of phase billing percentages due to new drafting production techniques employed by the firm which tend to cause more work to be done in Schematic and Design Development phases. The revised breakdown is:

Phase	Standard AIA, %	Revised, %
Schematics	15	20
Design development	20	25
Working drawing/specs	40	30
Bidding/negotiation	5	5
Construction administration	20	20

This revised breakdown of percentages, if implemented, can have a decided impact on cash flow and profitability. Systems Drafting allows a firm to do a better job for the client while at the same time improving the firm's financial base. Another bonus for the firm using Systems.

To further substantiate the fact that Systems does reduce working drawing production time, let's consider a report furnished to top management of Gresham, Smith and Partners which clearly indicates the time required for UNIGRAFS or systems produced drawings versus conventional drafting on past hospital projects. The projects are relatively equal in size, cost, and number of beds. (Cost differences reflect cost of labor and materials, seismic requirements and inflation for different locations and year built.)

Project	Project cost	No. beds	Total hrs. reqd. for working drawings
Carthage, Tenn.	$ 2,115,995	66	962
SPARTA, TENN.*	$ 2,083,365	60	608
Frankfort, Ky.	$ 5,389,489	122	2909
MYRTLE BEACH, FLA.*	$ 6,837,428	124	2237
Arlington, Tex.	$ 3,417,846	131	2560
PAYSON, UTAH*	$ 4,500,000	130	1699
Carlsbad, N.M.	$ 6,327,122	134	3066
HELENA, ARK.*	$ 3,688,972	131	801
Mary Elizabeth	$ 5,862,393	143	1580
HUNTSVILLE, TX.*	$ 6,563,000	140	1857

(Continued on page 8.)

Project	Project cost	No. beds	Total hrs. reqd. for working drawings
Chippenham 6-7 Fls. Renov./ Add.	$ 2,539,039	—	1007
N. FLA. REG.5-6-7 FLS.* Renov./Add.	$ 4,340,700	150	437
Lewis-Gale	$13,510,332	322	6711
TERRE HAUTE*	$13,059,000	284	3958

*UNIGRAFS or Systems-produced projects

The evidence that Systems pay is clearly shown in the above figures. Renovation and addition type projects can produce even greater savings because existing drawings can readily become part of the new drawing set. (Existing drawing sheets are photographically made into base sheets for new construction.) This is indicated by the difference in time required to produce Chippenham versus North Florida Regional Hospitals.

During the next decade, offices who develop an action plan for implementing systems which produces a stable chemistry of people, procedures, and equipment will increase in size and profitability. Those who do not plan and adopt new methods stand a good chance of not surviving through the decade.

Gresham, Smith and Partners made the commitment to Systems Drafting in 1975. In that year the firm had less than 50 employees in their main and only office in Nashville, Tennessee. Today the number of employees is 275 in their home office and four other states. There is a parallel between the use of Systems, which allows working drawings for hospitals to be drawn and released to contractors in 90 calendar days, and the growth of the firm.

Fig. 1-1. This text tells a major success story with systems. It's from the preface to the systems manual *Unigraphs—Unique Graphics for Architects and Engineers* by Edgar Powers, Jr., published by one of the leading systems firms in the nation, Gresham, Smith and Partners in Nashville.

2

EDUCATING YOURSELF AND STAFF MEMBERS IN SYSTEMS

UNDERSTANDING AND AVOIDING THE PEOPLE PROBLEMS

"What works best in teaching advanced technology and methodology?" The question is on the minds of many design professionals these days. Besides having to learn new technology, they have to turn around and teach it to their colleagues and staff. It's not easy. Conveying ideas from one mind to another requires extraordinary thought and skill. Run-of-the-mill educators can get away with giving students clichés, generalities, and questions without answers, but professional education requires far more. It requires that a maximum of solid information be conveyed in a minimum amount of time. That means data have to be organized, clearly stated, and presented in such a way as to anticipate and prevent misunderstandings or resistance to the data.

THE CYCLE OF IGNORANCE, FEAR, AND RESISTANCE

If you're thinking about educational methods for yourself or your staff and associates, remember that the first barrier to learning is fear. There's resistance to new systems. There has to be. People are worried about their jobs. They're worried about being relegated to the role of a factory worker who cranks out copies of pieces of buildings like so many machine parts.

Older design and drafting staff worry about starting over and having to learn new skills—skills which younger people will learn just as quickly as they can—maybe faster. That means more mature staff lose the benefit of years of experience. Their hard-won skills will suddenly decline in value. That's a valid cause for fear, and it has to be acknowledged and handled sensitively. If you think you can do without sensitivity in dealing with people's fears about new systems technology, you may as well plan on doing without the systems.

All the resistance to systems is born of fear, and the fear is partly justified and partly born of lack of knowledge. Now we have to confront a subtle problem that can undercut any effort at continuing professional education. People start with a knowledge gap. The continuing education we're talking about is designed to fill that gap. But people cannot possibly learn well or use what they learn if they think the whole process will profit somebody else at their expense.

You'll have resistance born of fear, fear born of ignorance. When the resistance hits head-on with management demands, the fears get worse. The main way to break through resistance is through knowledge, but, as the Catch-22 here, just as it takes learning to cut through fear and resistance, that same fear and resistance are blocking concerted focus on learning.

Of course you can tell the troops to shut up and just do as they're told, but you'll get nowhere fast—especially in a self-motivated activity like learning. That approach has its place, but it won't work for you as general policy. The hard-nosed approach to employee resistance will gain a superficial "OK, we'll do as we're told" acquiescence from the staff, but production will hit rock bottom. Take that from me personally. I've seen it not once, but dozens of times.

You can break through the self-reinforcing cycle of ignorance, fear, and resistance more easily than you think. The tool that cuts through it all is plain old human sensitivity and understanding. These qualities clarify people's concerns, and that's what we're after, clarity. Most employees don't expect you to agree with their concerns. All they want to know is that those concerns have been heard, understood, and considered. That's all you or I want in any conflict with another. We're mature enough to know we can't always get agreement. But we want our concerns, doubts, reservations, and fears to be recognized and respected.

BE CANDID AND SYMPATHETIC

Instead of pretending that people should immediately, willingly latch on to new techniques because the company says so, just let them say out loud what's on their minds. Let them say it even if it sounds like they're disagreeing with you. And repeat what they say so they know that you know what they're worrying about. Then they'll start to drop the fear and resentment over the impression that something unpleasant is being rammed down their throats.

Keep in mind too that the first efforts at any new physical action—whether paste-ups, appliqués, ink or freehand drafting—all will be somewhat awkward at first. That slows people down. Their first efforts may be slower and of lower quality at the same time that the office is making noises about how everybody should be speeding up and producing higher-quality work. It may not work that way, and the notion that everything should suddenly speed up lays an impossible burden on the staff.

Not that there can't be notable first-time results such as I've already described. But those are achieved by motivated people who aren't afraid of the systems. We're talking here about the pitfalls of fear and resistance. It won't always be there. It may seem minor. But it can catch you when least expected.

REPEAT INFORMATION

It is a major fact in education that people only rarely grasp an idea the first time they hear it. They may say they do and pretend they do, but they don't. So if you teach something new and think the job's over, you'll be unnecessarily frustrated. You've just made the first step. Also, don't think that people are listening even if they're looking intently at you and nodding. People only listen part time. You'll see that at any lecture when people ask questions which have already been clearly answered in the talk. The questioners just happened to be mentally out to lunch for a few moments when the topic was covered. Why not? We all do it.

All this is natural and normal. If you accept the fact that people don't listen and won't know what you're talking about at first even when they give every indication that they do, then you can allow for it and work with it.

Nobody actually knows something until he or she can create the words to explain it to somebody else. So do what's necessary to get those you teach to act as tutors to others. When they can teach it, they've learned it.

ENCOURAGE DISCUSSION

It's also a fact that it's virtually impossible to be unconscious while asking a question or responding to one. Thus, in an educational setting you want to do all you can to encourage and reward questions. First, you say out loud that you want inquiries and that there's no such thing as a stupid question. Any question is a request for data, and that's what education is all about—the transfer of data. State this a number of times, and people will begin to hear it. Then prove it by rewarding anyone who asks a question. Even if it's the dumbest question you've ever heard in your life. You answer it and then specifically thank the questioner for raising the issue. A warning here: Only answer the question. It's fun to go further and talk a bit about things the question reminds you of. That'll bore everyone to tears, and it does not reward the questioners.

STRUCTURE THE INSTRUCTIONS

Besides being candid about those unspoken concerns and rewarding people who ask questions, as the next key tool in educating staff to new techniques you should follow a strictly logical progression of data. Clarity is the standard of good educational communication.

That doesn't mean it has to be slow or plodding—just clear. That means not using technical terms without defining them. It means proceeding from what people know to what they don't know with generous use of comparisons, examples, and metaphors to help them link up the new ideas with old ones. It means dealing with misunderstandings and obsolete data. Everybody has obsolete ideas about systems, what they are, how they work, and what their problems are. So you express those common misunderstandings out loud. Let others say them too. Make sure they get a thorough airing, or people will still carry them as part of their mental baggage.

EASING INTO SYSTEMS

I'll close this section with an outline structure for presenting information on systems graphics to your staff. If you follow this structure in training your people—or yourself—as well as follow it as a general plan of implementation, the whole process will go remarkably smoothly.

I call it the five stages of transition. You may recognize it as the organizing principle behind my book *Systems Drafting*. It starts very simply with little or no investment and builds from there. The preliminary steps save a remarkable amount of time and money in themselves and don't really threaten people. They make it possible to move successfully to the next stages, and the stages that follow will enhance the value of the early stages.

STAGE ONE— HOUSECLEANING

"Housecleaning" means reviewing your documents for overdrawing, redundancy, oversizing, etc. Identify what is being drawn that doesn't need to be drawn, and eliminate those items from future documents. It also means identifying the most common repeat items used from job to job: symbols, north arrows, drawing titles, general notes, construction details, etc. Plan on creating standard elements that'll be copied by machine in the future instead of by hand drafting.

Since you're looking forward to copying and reusing data, your originals have to be extraordinarily clear, crisp, sharp, and opaque. If they aren't, they won't reproduce well and won't be reusable. It's simple to say this, but this is the first stumbling block in systems, and most offices trip right over it.

Housecleaning is an appropriate theme for in-house professional development seminars for drafting staff. Encourage staff suggestions regarding overdrawing and repetition.

Once staff have been trained in clean functional drafting, only then is it appropriate or profitable to train them in using the appliqués, light-table graphics, etc.

STAGE TWO—REUSING SIMPLE DATA

This means actually copying the repeat data you've identified in stage one and applying it in future projects. Standard or master details are a good focal point. Preprinted appliqués, use of the office copier, keynoting, commercial dry transfer and stickyback appliqués, translucent paste-ups, etc., are all parts of this stage of development.

Note that it won't pay to start reusing data if your graphics aren't simplified and extremely sharp to begin with. That's why this is the natural follow-through from stage one.

STAGE THREE—USING NEW TOOLS

At this point you exploit graphic tools and equipment to create extra clear and sharp originals and copies of originals. Make sophisticated use of office copiers, use the office typewriter and the word processor to do notation and titles, use cameras to record existing job data instead of measuring and drawing it, etc. Generous use of light tables for appliqué and paste-up work and a switch to the graphic artist's tools and equipment are the main theme of this stage.

"Quicky" office orientation sessions work well to acquaint staff with how to use new equipment. It's best not to overwhelm them with massive training on all new equipment and tools at one time. It's important to provide a brief formal introduction when new equipment is moved into the office. Otherwise the staff won't know if it's all right to use it or not, or what the limits on its use are, and the new capital investment may gather dust.

STAGE FOUR— COMPOSITE AND OVERLAY DRAFTING

At this point the office is well into a graphic arts approach to making documents. It's a very simple step now to proceed all the way into paste-up (also called "composite") drafting. This is where the most dramatic time and cost savings will occur, and this is the step that will integrate all the preceding materials, tools and techniques.

Once the essentials of paste-up drafting are understood, it's a natural step to proceed to the complexities of overlay drafting. Overlay pin register drafting can be uncomfortably complex to first-timers who have had no graphic experience whatsoever. The techniques and, most importantly, the theory behind them are much easier to grasp after experience with the simpler paste-up processes.

STAGE FIVE—TOTAL SYSTEMS DRAFTING

This stage follows rapidly after a little experience in stage four. It means the office has a grasp on all the technologies and can sensibly pick and choose among them for any particular project and any particular drawing.

Stage five offices are characteristically fully into multicolor offset printing of their working drawings, computer-aided design and automated drafting, and various hybrid innovations of their own. It's the inevitable final result of a complete exploration of systems drafting. In fact it's the inevitable final result of any long-term design and working-drawing quality control system.

If you follow the five stages structure in training your staff as well as in putting the systems to work, you'll fill their knowledge gap in reasonably palatable doses. Staff will learn best by proceeding from the simple to the complex in a logical sequence, and they'll be reassured. They'll see that changes don't hurt. They'll see that their fears aren't realized. They'll see that their work becomes faster-paced, more interesting, and more rewarding on every level.

Here are two examples of firms that follow a rational and systematic approach to staff training. One outstanding example is Stone Marricini and Patterson Architects and Planners in San Francisco. Morry Wexler, production coordinator, conducts periodic workshops for all staff using an abundance of well-conceived, humorous display boards. That's for starters. To really make it work, he gathers together newly formed teams assigned to new projects and goes through the training again. He works with the project managers in deciding how different drawings are to be handled: paste-up, photodrafting, photocopy, or overlay. Unlike many drafting room managers, he doesn't toss systems at the staff and leave them to sink or swim. He's there and always ready to help with understanding and making best use of the new technology.

In a related vein, Jim McGrath and Colleen Hyde of John Graham and Company in Seattle have initiated an officewide communications system. It's a newsletter that describes techniques, the new tools and materials to use in the graphic work area, and how to use them. It's made clear that people are supposed to take advantage of these resources. The regular humorous newsletter helps cut through the hesitancy and ignorance that commonly block people from putting it all to work.

Finally, just for yourself, remember that most of learning any new subject is learning the vocabulary and identifying what the words actually refer to. And then for complete understanding of what's been learned—get in some hands-on practice. Experiment. Make mistakes on a small scale where it doesn't matter. Encourage others to do the same.

3

TRANSLUCENT PASTE-UP DRAFTING

THE ALL-PURPOSE, LOW-COST REPRO SYSTEM

There's no way I could overstate or exaggerate the versatility and value of the technique you're about to see demonstrated in this section of the book. It's an advanced form of the general composite drafting techniques shown in my earlier book *Systems Drafting*. "Advanced" in that it is inexpensive, convenient, easy to learn, and achievable with equipment and materials already on hand in the drafting room. And it'll save you more time and money in more ways than any technique or tool described anywhere in this book.

Let's make clear why this is so. Then we'll show the technique in action through a series of step-by-step "how-to" photographs.

All the repetitive drafting elements we mentioned in the previous chapter can be handled as paste-ups. If you do a site plan, for example, the plan is most convenient and economical to create and revise if it's a paste-up. The same is true with floor plans, furniture layouts, renderings, elevations, detail and schedule sheets, and every kind of engineering drawing.

Most people in the field can't buy this idea at first. The reason is that when you say "paste-up" (or "appliqué" or "composite"), people automatically think of opaque paste-ups that have to be sent out to a repro shop to be photographed and processed on a high-priced silver emulsion pho-

to-washoff sheet. Sending paste-ups out of the office creates scheduling problems. And budget problems.

If you tell people they can make *translucent* paste-ups in the office to print by diazo printing, they have another problem. Most people who have tried using stickybacks extensively or homemade paste-ups find they cannot print them successfully on the diazo machine. The prints are full of ghosts, shadows, tape marks, overlap marks, and the line work prints are broken and fuzzy at the same time. So to most people, "paste-up" conjures up associations with outside repro services, inadequate results, inconvenience, and high costs.

THE SECRET BEHIND MAKING INEXPENSIVE, GHOST-FREE PASTE-UPS

Here's how you can do it cheaply, in-house, and get good results. First, there's a trick to it. The trick is not to use original drawing elements as part of your paste-up. Take every original item created for a drawing and copy it into a form that is intensified, denser, and more reproducible than the original.

Suppose you're doing some floor plans of a housing project. Take the original drawings of repetitive apartment units and copy them in multiple on translucent diazo film. Then assemble those units onto a clear polyester carrying sheet and adhere them with transparent tape. You now have a translucent paste-up. You can treat that paste-up as a tracing and, with care, run it through a diazo machine and make a reproducible print that becomes the equivalent of a final tracing and can be revised, added to, etc. A paste-up of work like this goes not just 30 to 40 percent faster, but two to three times faster than traditional drafting.

But how about the ghosting when you make a diazo reproducible? This is at the heart of it all. The copies you make for paste-up have to be intensified. The lines have to be as opaque as possible, more so than graphite lines on paper can be. So, if you're starting with graphite and tracing paper originals, you have to translate that into something better. That "something better" is, in all the examples to follow, a product called "reproducible diazo sepia line polyester film."

Here's what those terms mean and why they're so important:

1. "Reproducible" means a print from which you can make other prints. In standard architectural and engineering diazo printing, the "paper sepia" is well known as a reproducible medium. Other words that mean much the same thing are "intermediate print," "second original," and "transparency."

2. "Diazo" means the commonly used blue line or black line or sepia line printing process used to make regular architectural and engineering prints from tracings. It's the process that has replaced blueprinting. It requires an ultraviolet light source and an ammonia print developer. (Nonammonia replacements for diazo printing will not work well for the processes we're describing.)

3. "Sepia line" means a brown or reddish tone such as that seen on what are often called "paper sepias" or "brown line" prints. The sepia color has a yellow dye base which is especially opaque and resistant to ultraviolet light. It's often called "intensifying film" for that reason, in that you can make a sepia line copy of a drawing, and the sepia line reproducible print will be more intense, more opaque to the printer light than the original line could ever be. It'll be more opaque to ultraviolet light even if it looks translucent in regular light.

4. "Polyester film" means the plastic film popularly known by the Du Pont trade name "Mylar." Mylar is used for drafting films and print films, both the diazo type and the photo-washoff silver emulsion types.

The sepia line polyester reproducible film we're describing comes in varying thicknesses, measured in mils. One mil and 1½ mils are very thin, while 3 mils is an average thickness for most paste-up purposes and for making medium-sized reproducibles.

The sepia polyester film comes with a choice of a matte drafting surface on one or both sides. Double-sided matte surfacing creates problems in reproduction, so a matte surface on one side only is recommended. It also comes as a non-matte surface "slick" which has special applications in overlay drafting.

The emulsion, which appears as brown or red after exposure to ammonia fumes, may be erasable, nonerasable, or removable with liquid eradicating fluid. Most experienced drafting staff have had memorably bad experiences with the old-time paper sepia products. Those sepias could only be changed with eradicating fluid which stank, was caustic, and would leave the reproducible sheets looking like some skin disease. Those days are behind us.

Now the plot thickens. The importance of this product—sepia line diazo polyester reproducible film—is that if you transfer drawings, notes, photos, etc., onto such film, you can make a paste-up from which you can make a ghost-free reproducible print.

HOW TO DO IT

Here's the process:

1. Create or obtain the original.
2. Make a copy or copies on sepia line diazo film.
3. Use the sepia film copies for paste-up on a transparent polyester carrying sheet.
4. Copy the paste-up of sepias onto the same medium, that is, print the paste-up onto a sepia line diazo polyester film reproducible.

How do you get a ghost-free reproducible print of the paste-up out of this? That comes from the character of the sepia line film. Since the sepia emulsion lines of your paste-up elements are especially resistant to the ultraviolet light in the print machine, you can expose the paste-up to that light long enough to burn away any traces of tape marks, paste-up edges, overlaps, white spots from cutouts or erasures, etc.

There are some further rules to make this work. First of all, the original line work you reproduce for paste-ups must be dark enough to reproduce clearly in the first place. You can't intensify faded or broken lines.

The first-stage reproducibles on sepia line film have to be reasonably free of background haze. That'll require that you not try to rush the printing process. Otherwise, the time it takes to burn out hazing during the second stage of printing may be enough to start burning out line work. My experiments indicate that for most paste-up purposes, you should use 3-mil sepia copies with a matte surface on the upward side and emulsion on the back.

The sepia line diazo polyester film you use must be of good quality. There are what are called "garage coaters" or "backyard coaters" in the diazo products industry. They produce low-cost sepia films that are fine for very short-term use, but they'll print foggy to begin with or fog up on you very quickly after printing. Sources of the best-known and most reliable of sepia polyester films are listed in the *The Guidelines Reprographic and Computer Resources Guide* cited in the back of this book. Widely used sources include Ozalid (formerly GAF), James River Graphics, Precision Coatings, K&E, and Teledyne Post.

All diazo print products have to be well-protected from light and air, and they have a limited shelf life in any case. That creates inventory problems and occasional unpleasant surprises, but it's a minor price to pay for the advantages.

Translucent paste-up drafting makes it possible for you to do photo drawings, typewritten notation, appliqué drafting, screened background sheet printing, and small- and large-scale paste-up with great ease and convenience. For all-out convenience, you'll need a vacuum frame, flatbed contact printer as described in the illustrations, but you can experiment and do a remarkable amount of small-project work with these systems with just a plain old tabletop rotary diazo printer and developer.

OTHER WAYS TO GET SIMILAR BENEFITS

Many architectural and engineering firms have dry toner, photo-quality office copiers that will copy reproducible images, very opaquely, on transparent polyester film. These firms make their repeat units on these films and use them for paste-up instead of using the diazo sepia polyester. If you have the equipment and don't have a problem with possibly flawed 1 to 1 size fidelity, this method is convenient and faster in the initial stages of making multiple copies for paste-up purposes.

Newer office copiers provide 50 percent reduction capability which greatly augments their value in systems for architectural and engineering applications. And there are the large-size enlargement-reduction copiers[1] which will copy from opaque paste-ups and convert them into tracing paper or polyester reproducibles. Many firms swear by such equipment and either maintain it in house or use it at the repro house as a low-cost alternative to photo-blowback and photo-washoff reproduction.

[1] These include the Xerox 2080, Xerox 2020, and Océ 7200. Other brands and models can be expected in 1983–1984.

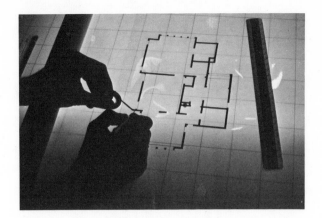

Fig. 3-1.   A drafter prepares a ¼-inch-scale house plan with graphic tape. The tape is applied on a transparent polyester (Mylar) carrying sheet. A lightly sketched schematic plan is used as a tracing underlay to guide layout of the taped walls.

It's hard to read through black graphic tape to see openings for doors and windows. For that reason, many offices prefer dark, ruby red litho tape. It's opaque to re-pro light but translucent for convenient cutting.

Fig. 3-2.   Tape drafting *is* faster than hand drafting when dealing with thick lines or complex dot-dash lines. It takes some practice to bring the speed up, and drafters have to learn not to slow themselves down by cutting tiny wall pieces such as nubs for door jambs. Small items like that should be drawn with ink instead. (Most ink for technical pens will adhere to clear polyester, surprisingly enough. The inking has to be done carefully though.)

When you have to indicate different types of wall construction, you can use different patterns of graphic tape. Or use wall key symbols and a wall construction schedule.

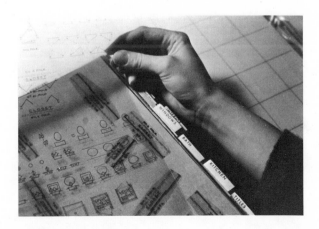

Fig. 3-3.   This is a three-ring binder "master file" that contains many standard elements of residential construction. It has index tabs labeled "Doors," "Windows," "Closets," "Room Names," etc. This office does lots of housing, and the staff realized long ago that most housing contains exactly the same elements over and over. They've created those elements by drawing them individually on tracing paper and copying them on sepia line diazo reproducible film. A graphic work assistant cuts out the elements and places them in these three-ring binders for all the design and drafting staff.

Fig. 3-4. The "bedroom" name is applied to the plan. Small elements like these are usually adhered with small pieces of transparent tape. Some offices use double-sided sticky tape. Some use waxy "glu-stick" products from art supply stores. The main criterion is that the adhesive be reasonably translucent and repositionable.

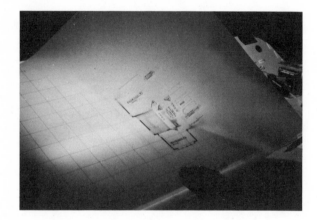

Fig. 3-5. Here, you see that the added room names, doors, etc., were applied on an overlay sheet rather than on the original taped floor plan.

The first rule of systems drafting and systems graphics is to create all original work in such a way as to be able to reuse it if and when the need comes up. Therefore, floor plans are done without adding room names directly to the plan. That way, if the plan is reversed or flip-flopped or used upside down, nothing has to be erased. All rendered elements are done on overlay, as are titles or optional variations in the design.

The original floor plan is kept reusable from design development into working drawings. Now the staff can go a step further. If the variable data such as notes, dimensions, etc., are done on overlay, the plain floor plan can be screened as a shadow print and combined with roof framing, the foundation plan, furniture plans, etc., without having to be redrawn. (Such screened shadow prints can be made with diazo printing as you'll see in the section on overlay print processes.)

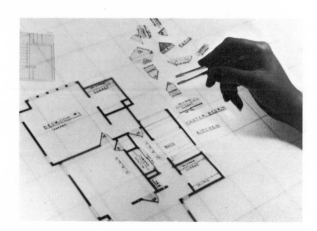

Fig. 3-6. Despite the small size of these repetitive elements, many firms find it fully practical to print, store, and assemble them on working-drawing paste-up sheets. Although it can be said that it's virtually as fast to draw a door swing for example as to stick one on with tape or glu-stick, the paste-up method is actually faster and less laborious. It's especially faster when it comes time to make changes. Then it's just a matter of picking something up and moving it around as opposed to erasing and redrawing. That's where time savings show up most dramatically.

Fig. 3-7. We're going to see a multistory hotel created from just a few original pieces. First, this pair of guest rooms has been worked and reworked in tape and as a paste-up.

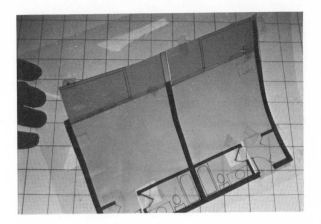

Fig. 3-8. The paste-up is laid face down on the emulsion side of a sheet of sepia line polyester. It's to be reproduced in exact 1 to 1 size, so photocopying or a rotary diazo print machine won't do. Those processes tend to affect the size of reproductions, and you'll get accumulated error in any paste-up process when doing the total building floor plan. For 1 to 1 reproduction, a vacuum frame is used to make contact prints.

Fig. 3-9. The floor plan guest rooms are being cloned. If you're rushed, you can make a copy from the original, then use those two to make two more, those four to make four more, etc. But you'll tend to lose some line quality in the process. The print media is 3-mil sepia line polyester. The emulsion is on the back side and is erasable. The top side is matte surface for any inked-in additions or repair work that may be required.

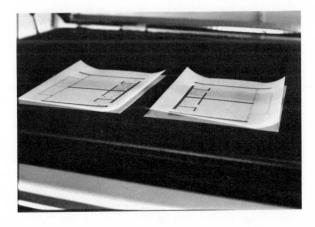

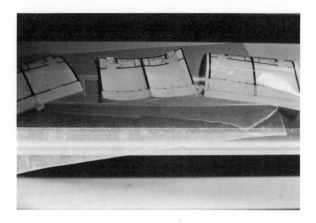

Fig. 3-10. After the guest rooms and sepia polyester print media are exposed to ultraviolet light on the vacuum frame, flatbed contact printer, the image is developed in the ammonia fumes of a regular rotary diazo machine.

Fig. 3-11. The paste-up process begins. Paste-up at this stage of design development follows the schematics laid out by the designer or design team. All these data are created as base sheet information, and variable data such as room names, notes, furniture, etc., will be done on overlays.

Please note that the grid lines on the light table are for alignment and guideline purposes only; they do not reflect the structural grid of the building.

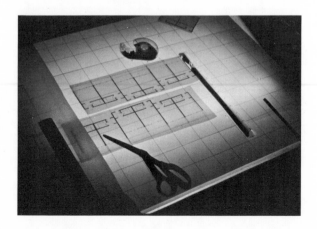

Fig. 3-12. The first cluster of rooms is roughed into place with drafting tape and then adhered with transparent tape. That cluster of six rooms is then copied on the vacuum frame to make the second group of rooms. All is taped onto transparent polyester.

Fig. 3-13. Besides guest rooms, the hotel floor plan requires elevators, exit stairs, service areas, etc. Following the rule that original data are not to be used up, these data are created on tracing paper or polyester, then copied.

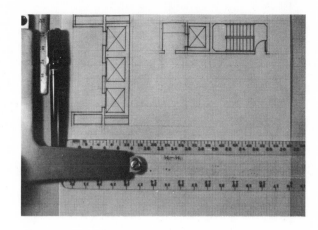

Fig. 3-14. This shows sepia line polyester copies of the stairs, elevators, etc., beside the original. The original is kept unused along with other original data such as room names, blocks of notation, etc. Now guest rooms and other graphic data are ready for the paste-up process.

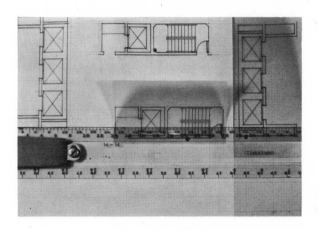

Fig. 3-15. Here's half of the building floor plan. Three pairs of guest rooms are taped in position across the upper part of the carrying sheet. That is matched with its mirror image copy made on the vacuum frame. Ordinary utility-grade transparent adhesive tape or double-sided tape works well for these paste-ups. The tape should be strong enough that it doesn't tear apart in small pieces when it has to be removed.

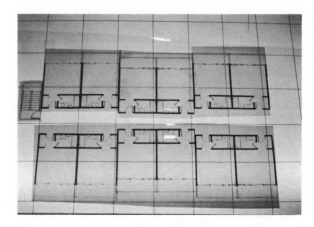

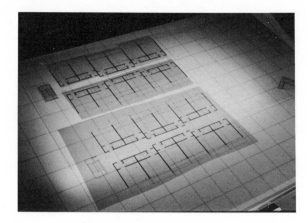

Fig. 3-16. Just as the first three pairs of guest rooms were copied to complete one wing of the building, so that whole wing is copied in turn on the vacuum frame to make the other wing.

Connective line work such as the perimeter of the wall or windows at the end hall exit stairs is left for the final reproducible.

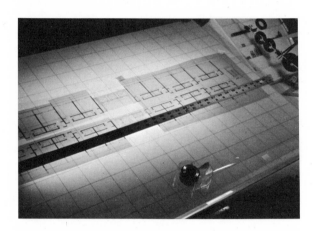

Fig. 3-17. The connective bridge of elevators and stairs is added and the second half of the building taped into place. The whole process you're seeing here took place between the end of lunch and 4:00 p.m. The method is extraordinarily fast for this type of building, and for hospitals, jails, motels, office buildings, etc. And changes are remarkably fast to make too. One caveat though: When copying any repeat element for extensive paste-up, make sure the original element, like these guest rooms, is exactly what you want.

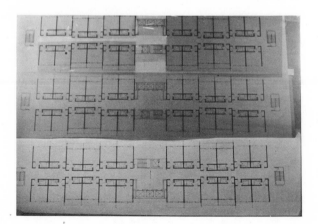

Fig. 3-18. A combination of the original paste-up across the top, a sepia polyester reproducible of that paste-up across the middle, and a blue-line print from the reproducible across the bottom.

This illustrates the steps in reproduction and, by implication, shows how you can get multiple floors of a multi-story building—a large amount of drafting—with virtually no drafting at all. And all by one person within less than an afternoon.

Fig. 3-19. The hotel project continues with the reflected ceiling plan. Ceiling grids are a constant—they're ubiquitous in type and grid sizes. So rather than being drawn over and over, they're made once and then copied on sepia line polyester reproducible media for ongoing use, project after project.

Here a portion of translucent grid sheet is laid over the floor plan, positioned, and marked for cutout. It's far faster and easier to position and reposition any grid in this fashion than to start drawing one, find it's not correct, and have to redraw it.

Fig. 3-20. The grid is cut along the marked tick marks or lines. A no-print blue pencil is used to mark cut boundaries.

All such work must be done on a light table with a cutting surface and with a grid for visual alignment of all the paste-up elements.

Fig. 3-21. The grid cutouts are set in place. They'll be rough-taped in position first with drafting tape and later adhered with transparent tape when positioning is checked and confirmed.

Fig. 3-22. All this paste-up work is *not* done directly on a floor plan sheet. It's done on a transparent polyester overlay sheet. The two images—floor plan (screened) and the reflected ceiling plan—will be combined as a base sheet for the electrical engineer. The electrical work will be created with ink, tape, appliqués, and paste-up on its own overlay.

If and when the floor plan is changed, only the paste-up elements on the overlay will have to be changed to match. In traditional drafting, the reflected ceiling plan and floor plan would be drawn on a single sheet. When changes in plan were made, that plan had to be erased and redrawn to match along with the ceiling data. Now changes need be made only on one original, not on all the hand-drawn copies commonly made in traditional practice.

Fig. 3-23. This particular sequence is a hybrid. It's a combination of paste-up drafting on an overlay sheet. Now we're adding another systems element to the package—appliqué drafting.

Fig. 3-24. Many architects and engineers now have small elements that are used over and over again, such as these light fixtures, printed on die-cut acetate stickyback films. Since they're precut, you only have to peel them off individually and stick them in position. If there's a change later, they're easily repositioned. Most of the applique products companies can make such custom sheets for you for quantity printing.

Fig. 3-25. The final result of a combination of techniques: paste-up for the floor plan base sheet, paste-up for the overlay, preprinted ceiling grid sheets, preprinted stickyback light fixture elements. All combine to create an easy to reproduce original that can be revised as easily as it was created, without drafting and without erasing.

Fig. 3-26. Now we'll see the translucent paste-up and reproducible process used for making exterior elevations. In this example the client needed to see several variations of interior courtyard designs for renovation within an historic and design-regulated part of the city.

After sketching some options, the designer now drafts the pieces of the facade design accurately and to scale.

Fig. 3-27. The original tracing is reproduced in multiple. Various possible components are created and then copied onto sepia line diazo polyester.

Fig. 3-28. Lots of copies of the repetitive elements are made. And good copies can be used in turn to clone more copies.

Fig. 3-29. The facade elements are pasted up to compose four sides of the interior courtyard design. Several other variations of design are also assembled.

In this particular case, the architect reports that they had a conference room full of design options to review with the client, all created within very little time. The sepia line polyester film used was of extra high quality emulsion to ensure that all the fine line work of the ornamental metal would reproduce well.

(Courtesy of Parez Associates)

Fig. 3-30. Now a replay of the creation of building facades. First a portion of repetitive fenestration is drawn on tracing paper, or drawn on polyester with ink. That image is intensified by copying it onto the sepia line film. It's copied in multiple, and for exact 1 to 1 size fidelity, it's copied on a vacuum frame.

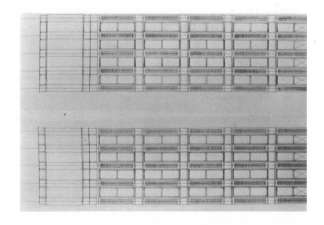

Fig. 3-31.   The repetitive fenestration elements are taped to a transparent polyester carrying sheet, and then that whole block of the building is copied in turn to make a new reproducible.

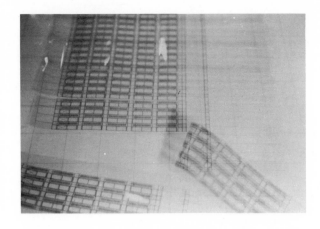

Fig. 3-32.   A final paste-up is made combining the building elevation with the lower-story construction, notes, titles, etc. Reproducibles are similarly made for the other elevation views of the building.

(Courtesy of John Graham & Co.)

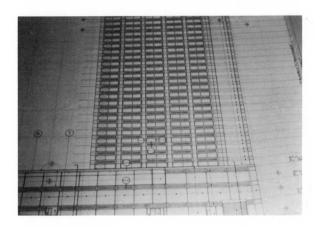

Fig. 3-33.   Lettering machines are improving all the time, and every design office should have one for doing titles. They create good reproducibles of near typeset quality and can be used by anyone, so your higher-paid design staff doesn't have to spend time doing rub ons.

Fig. 3-34. Since the lettering machine–printed images are easily scratched and damaged, it's best to translate the titles into a more durable form. Also, it's wasteful to have staff redo the same common titles over and over again. So, as in this case, the office has a "catchall" sheet made of common titles.

Fig. 3-35. The catchall sheet shows frequently used drawing titles. Other sheets can include common detail titles, site plan component titles, etc.

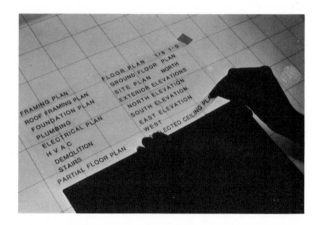

Fig. 3-36. The catchall sheets with lettering machine titles are copied in multiple onto sepia line diazo polyester. They can be printed either on the vacuum frame or in a regular rotary diazo machine.

Fig. 3-37. The graphic work assistant chops out the commonly used titles. Now that they're in sepia polyester film, they'll be resistant to damage, will print clearly when used on paste-ups, and are reusable.

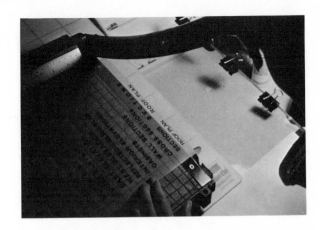

Fig. 3-38. The graphic work assistant sees to it that every design and drafting staff member has a complete set of titles for ongoing use. Here, they're stored in slotted vinyl sheets such as are used to store 35-mm photographic slides. The vinyl-pocket sheets come three-hole-punched to fit standard three-ring binders.

Fig. 3-39. Large rub-on, dry-transfer letters, numbers, and symbols are often hard to apply accurately or hard to apply without damage. This firm takes sheets of favored type styles and copies them on the vacuum frame onto sepia line film.

Fig. 3-40.   The sepia diazo film copies of the letters are cut out for filing by the graphic work assistant.

Fig. 3-41.   The letters are stored in three-ring binder vinyl-pocket sheets which are distributed to all design and drafting staff.

Whenever staff members have to do large-letter title or cover sheets, they pull the needed letters and tape them into position with transparent tape onto a clear carrying strip. Then they make a reproducible of the paste-up to become a paste-up element in turn on the final carrying sheet. The letters are then pulled off the original paste-up and reinserted in the vinyl-pocket sheets. It's extremely fast and easy to position and tape down letters in this form compared with dry transfer.

Fig. 3-42.   There is a wealth of ready-made graphics that can be converted from opaque form to transparency on the plain-paper copier and transformed again into very intense line copies by sepia line film printing. Here, some tracing paper is assembled to run through the office plain-paper copier.

Fig. 3-43.   The opaque originals are placed on the copy bed. Architects and engineers find data such as building code sections, elements from old prints, road maps (for vicinity maps), photos, rendering entourage as from Ernest Burden's book of trees, cars, people, etc., and copy them onto clear films for cutout, filing, and later paste-up.

Fig. 3-44.   In this example, images of trees used in drawing site plan presentation drawings are first copied onto tracing paper to translate the opaque image into a translucent one.

Fig. 3-45.   Many designers now produce complete rendered presentation drawings by doing paste-ups of rendering components such as trees, shadows, material textures, human figures, etc., on overlay sheets atop plain line drawings of design development plans and elevations. That leaves the plans and elevations clean and clear for potential direct reuse in working drawings.

Fig. 3-46.   Some firms that have had trouble producing good-quality renderings and presentation drawings find they can assemble such ready-made components into thoroughly acceptable presentations very quickly. They collect them from every source they can find: books, magazines, old office renderings, etc.

Fig. 3-47.   Another view, showing exterior elevation entourage created with preprinted paste-up trees, cars, and people.
(Courtesy of Morry Wexler, Stone Marricini and Patterson.)

4

INNOVATIONS, ADVANCES, AND HORROR STORIES

CROSS-FERTILIZATION OF IDEAS

"Please don't publish what we're doing in the drafting room," said the production coordinator in a California architectural office. He had given me some valuable tips during a telephone interview I was conducting, and now he didn't want others to know about the firm's achievements. It wasn't the first time this had happened.

Some pioneering and innovative offices have achieved an exceptional competitive edge over other firms, and they want to hang on to it. "We've worked hard to make our system work," said an architect at an American Institute of Architects (AIA) convention at which I was lecturing. "We don't feel we should just give all that away."

Most innovative offices aren't secretive, but like the fellow mentioned above they usually don't just give information away either. Pioneers in reprographics provide paid consulting services for other offices. Others trade ideas and information with like-minded colleagues. Some groups regularly meet for lunch or breakfast sessions.

One product of such luncheon meetings is the Northern California AIA (NCAIA) Task Force on Production Office Procedures. Members still meet to trade notes informally, but they also formally shared much of their work with other professionals by publishing an outstanding systems-oriented manual on production office procedures. Contact the NCAIA office in San Francisco or refer to the *Resources Guide* cited at the back of this book for current availability and price of the *Production Office Procedures* manual.

Such groups also often have become the nucleus for computer user groups. They trade information, have speakers at meetings, share software, and, in general, help smooth the way for each other in adapting new technology to architectural and engineering practice.

Here are some ideas that have been spawned and nurtured in idea trade sessions among systems-oriented architects, engineers, and designers:

• A prize-winning building designer brought up an important question: How could he do more design studies and refinements in less time without interfering with the flow of the design process? Could reprographic systems help?

There is an easy two-part answer: First, composite systems (paste-up graphics) would let him quickly create varied plans and elevations without extensive drawing and redrawing. Second, if he used basic plan and elevation configurations on underlay base sheets and did variations and refinements on transparent polyester overlays, he could avoid repeated drawing and redrawing of the basic, relatively unchanging building elements.

He discovered other design uses for base sheets and transparent overlays. Besides re-studying design components, he could study and restudy elements of entourage, shadow, and texture for final presentation drawings. Since those variables, including presentation titles and symbols, were on overlays, the basic final building configurations remained "clean" for direct use in working drawings. Most impressive to the designer was the fact that his design intent survived with minimal distortion during the transition to production drawings.

- A housing specialist hadn't gone far enough in his thinking about systems. He thought he had already systematized everything possible in his repetitive building types. He dropped in on an informal AIA chapter systems seminar and found he was still doing unnecessary original work on each project.

Keynoting, for example, was a new discovery to him. That's where drawing notation is listed in a typewritten legend on one side of the drawing sheet. The notes are numbered, and it's the numbers which appear in the field of the drawing. It reduces all that repetitive notation lettering and can clean up and clarify a drawing considerably. It's also sometimes been overdone to the point of unreadability.

The housing architect was shown how he could use standard keynotes from project to project—on certain types of drawings. Drawings such as roof plans, exterior and interior elevations, site plans, foundation plans, framing plans, and building cross sections usually included many of the same notes, job after job. That meant he could make standard, preprinted keynotes on a numbered list that would apply to virtually every housing project.

The architect took off with the keynoting idea and simplified it. He typed columns of notes that usually went on various parts of foundation plans, exterior elevations, framing plans, cross sections, etc. He had the note columns printed onto thin-acetate, stickyback appliqués (the thin acetate avoids ghosting on diazo prints). Now he cuts out column lists of standard notes, applies them on original drawings, and draws in connecting arrows between notes and building components. He feels he has gained the economies of standard notation and keynoting without their disadvantages.

- A structural engineer had reached a point in life where he wanted to curtail his practice, stop managing staff, and just work part time at home. He saw an architect's presentation of paste-up drafting at a repro house–sponsored systems meeting and decided that might be just what he needed.

After some experimentation, the engineer discovered he could print his schedules, sections, and details sharply onto high-quality sepia intermediate print paper. This sepia paper is erasable and doesn't have the extreme foggy background that used to be common to the product. The sepia line reproduces well because, as described in connection with the diazo polyester print media, even though it's a copy, it's more intense than the original.

The engineer discovered that he could draw repetitive elements once only, copy them on the sepia paper in quantity, tape them on vellum carrying sheets, and run them through his small rotary diazo printer. That way he could make new sepia paper reproducibles of his paste-ups. It took practice to do this without slippage or damage to the paste-up. He quickly learned that any paste-up elements that go into the rollers of a print machine must be taped only at the top corners so they can flatten out as the papers pass through the print machine.

Print quality is not 100 percent crisp with this method, and there's some slight stretching and distortion of size of the paste-up elements when they're printed. And there's some ghosting of the paste-up elements on the reproducibles. Much of that ghosting disappears in reprinting, and the engineer says those appearance problems are architect worries, not engineers'.

After more practice he found out that a lot of new information created for a project was very often applicable to later jobs. So he taught himself the "golden rule of systems": "Don't use up original data. Save it for your data base and use copies."

The engineer now has his early semiretirement. He works at home and confided to me on my last trip to Florida that he earns more part time and alone than he ever made trying to run an office. And he's stopped taking downers.

- An interior designer in Manhattan likes to work directly with clients in creating room arrangements. He had all the furnishings and equipment he worked with printed on small pieces of cardboard. His staff, clients, and he would finalize plans by moving the little template pieces around building plans. Then, when plans were agreed upon, they would be redrawn, printed, and submitted to the client for final checking and approval.

After working with some systems-oriented architects, the designer had a brainstorm. He could have all those furnishings reprinted on clear film, move them around on the glass bed of a bottom-lighted, tube-type vacuum frame, and make print copies right on the spot.

He tested the idea by making sample plan elements and placing them on a floor plan tracing. The tracing had been laid atop a paper print sheet. After doing a quick trial plan, the designer laid a sheet of glass on top of the plan to flatten everything and keep it all in place. He left the combination of furniture pieces, floor plan tracing, and print sheet to "cook" under the office's fluorescent lights. After a half hour under the lights, the exposure was complete. He developed the print with ammonia, and it came out looking remarkably good.

He ordered his vacuum frame and now does design work by himself and with clients, working directly on the printer. When a plan looks right, he lays down a reproducible print sheet, pulls the lid, makes a print, and moves on to another variation. Many of his clients have had the experience of waiting days or weeks for prints from other designers that this fellow can produce on the spot. They say they've hired a genius.

EXPERIMENTS AND DISCOVERIES

One time-saver leads to another as offices pursue their own brands of systems drafting. One small-office architect began with a simple typewritten notation system. That led to a full-fledged changeover and the elimination of most of his drafting operations.

It started with a switch to "typewriter-size" drawings. The office does custom homes, renovations, and small commercial work. The staff found they could reduce drawing scales and readily fit jobs onto 11-inch by 17-inch (and sometimes even 8½-inch by 14-inch) sheets. Both fit into the standard 15-inch-carriage IBM Correcting Selectric II typewriter. (The 11-inch by 17-inch sheet requires a fold along the title block or general note border line in order to fit.) The small sheet sizes and typewriting both proved to be an enormous convenience in drafting.

The next discovery was the large-document copier (LDC). Small projects don't require many bid sets. So the office started using the LDC—a plain-paper copier that copies onto bond or tracing paper—to make reproducibles or job prints.

Then, two profitable realizations occurred: (1) Plain-paper copiers can copy on both sides of a sheet. Printing on both sides of job prints, as with magazine or book pages, doubles the effective sheet size and helps make up for the possible disadvantages of reduced originals. (2) Better yet, the architect found his copier would handle simple forms of composite, overlay, and appliqué drafting. With these techniques, plus typewriting and carefully planned reuse of design development drawings, most design drawing and production drafting in that office is now centered on inventive uses of the plain-paper copier.

Here's a capsule list of some of the best ideas, tools, and innovations architects and engineers are putting to work in the 1980s:

- They find extensive uses for National Aeronautics and Space Administration (NASA) aerial and satellite photos in site planning studies. NASA has 6 million images on file and provides transparencies or photos of virtually any place "at cost."

Specialized aerial and satellite photos are a boon for site selection, planning, and analysis. Developers and investors use them to spot opportunities that aren't visible at ground level. In one case an architect designing a marina and harbor complex obtained sequential photos of ocean tide conditions that proved priceless—at extremely low cost. Some site planners combine transparencies of site survey drawings with straight-down site photos at the same scale. The combined overlay image is a good check on survey accuracy and provides a site plan base sheet for design, presentation, and production drawings. Planners, of course, benefit from sequential films of urban and suburban development patterns, flood plain changes, varied seasonal views, etc. One problem arises when overlaying aerials with site surveys. There may be some distortion and mismatch because of the angle of photo image; therefore, don't expect to get a perfect match every time.

You can get a free data packet on Landsat, Skylab, and NASA aircraft photos (including information on how to order them) from the U.S. Geological Survey, EROS Data Center, Sioux Falls, SD 57198. Ask for the EROS Data Center booklet, the Selected Landsat Coverage Map, geographic order forms and inquiry forms, and related documents.

- Some architects are experimenting with moist-erasable felt-tip pens on polyester—a potentially faster, low-cost replacement for graphite or technical pen drawing. And several drafting teams draw multicolor line work originals in order to brighten and clarify their production graphics.
- A few firms have introduced "assembly notation"—a cross between drawing notation and specifications. The goal is to "verbally" describe construction in more depth on drawings and store the notes as adaptable, reusable standards for future projects. Assembly notes are stored in word processor or microcomputer disks and printed out with key numbers in legends keyed to the working-drawing plans, elevations, and sections.
- Architectural firms increasingly eliminate most dimensioning on their plan drawings. They rely on modular grids and identify wall locations and lengths by reference to modular grid coordinate numbers. Unique construction details are also numbered according to their location coordinate numbers.

PROBLEMS AND HORROR STORIES

About those horror stories we promised in the chapter title. Here are the most common snags people run into as they introduce new graphic systems, tools, and media.

MISUNDERSTOOD MEDIA

Although it's no major linguistic sin to refer to all print media as "paper," it does lead to problems. Print paper is paper; it's not acetate or polyester. Acetate or polyester is not paper; they're plastic films. When people order some "sepia paper," then, likely as not that's what they'll get instead of the polyester that they really wanted. And if they call for acetate, when they really mean polyester, they'll get an easily torn plastic film that has little of the durability and dimensional stability of polyester.

THE ABSENTEE MANAGER

Most everybody knows how to draft, and the process is so grindingly slow that not much damage is done if people operate under the wrong instructions for a few days. Not so with systems. It's a major change in habit which requires a radical change, and speedup, in thinking. So the managers who more or less toss some systems into the drafting room and then turn their backs on the process encounter the following situations:

- Drafters are instructed to use paste-up at every opportunity. The supervisor comes around later and finds someone pasting up individual dimension lines.
- Base sheets are distributed to drafting staff as background reference for their overlay sheets, and the supervisor later finds that drafting staff are tracing the base sheets onto their overlays.
- Staff are instructed to start writing out notation on check prints so that the typist can do final notation. Some drafting staff then letter their notes for the typist as neatly and carefully as if

they were hand-engraving a love poem on a titanium pinhead. Later they comment to the boss: "You know, this typewriting thing seems to take a lot longer than just doing it by hand."

• Someone hears about photodrafting just prior to starting a major building remodeling project. Photos would be a lifesaver in terms of beating production time and cost budgets. They would be, but by the time the instructions filter down to the drafters responsible for measuring and drawing up the existing building, they are not comprehensible. One staffer gets a government manual on photo preservation of old buildings, and they proceed to do exact-scale, distortion-free photos of the building facade. Then they put the photos on the drafting board and trace them.

• An architect tells her engineers to use keynoting on the current project. She later receives check prints and finds the engineering drafting staff has conscientiously listed and numbered every note in their work and then lettered all 670 of the notes onto the last sheet of their group of drawings.

THE PROBLEM WITH BELIEVING YOUR EYES

There's an old story about the fellow who returns home unexpectedly and finds his wife sleeping with his lifelong buddy. His friend says accusingly: "Well, what are you going to believe, your eyes or your best friend?"

A civil engineering firm observed that an architect with whom they worked had great success using the translucent paste-up system described in this book. They had earned some large fees that year and had no hesitation in splurging on a vacuum frame and a fully equipped graphic work center. Once set up, they were ready to use paste-up to the hilt on their very next job.

One problem arose. The firm ordered sepia polyester diazo film with which to copy repeat elements for paste-up, but one of the office principals observed that you could see through sepia-printed areas on the sepia film copies. "That's not dark enough to print well," he said. "Order something blacker."

The graphic work assistant knew that the sepia film images were somewhat translucent to the eye and looked inadequate. He also knew the developed sepia emulsion was perfectly opaque as far as the light in the print machine was concerned. And he knew that the "blacker" diazo films would print images that would look more opaque but were actually translucent to the print light. Thus, any paste-up made with the black line diazo film could not make a good reproducible print.

That knowledge was not put to use however. The boss who ordered "blacker" print media was one of those impatient types who cut people off when they try to explain a problem. Thus when he ordered something done and it was an error, the staff enjoyed going ahead and doing exactly what they were told.

The paste-up process proceeded, and the boss had some more suggestions. "If you're going to paste something up and it's not repetitive, then save the copying cost and just paste up the original," he said.

In short order the drafting room was filled with paste-up sheets which featured combinations of black-line polyester film, ink drawn on polyester, office copier stickybacks, and pencil on flimsy yellow tracing paper. When it was time for the first check prints, the prints came out splotchy. "That's OK for now, we just ran them too fast," said the boss. "The guys upstairs get splotchy check prints too, but the finals come out fine."

When it was time to make the first progress submittals, nobody could get rid of the splotches and shadows on the prints without washing out large parts of the drawing. Although the paste-up elements looked dark enough to reproduce—darker than sepia line paste-ups—many of them were transparent as far as the print machine was concerned.

The boss sent the whole paste-up collection to a reprographic shop. They photographed the sheets—no easy matter because of the extreme differences in line quality of paste-up elements. Fifty photo-washoff reproducibles were returned, at $105 each, and the boss said, "You know, this systems stuff is more trouble than it's worth."

In a similar vein, an architect had sent staffers out to a systems drafting workshop, and they returned raring to go. They moved right into a paste-up system and were doing everything exactly right and according to the rules. Only one problem. The boss didn't know the rules and could only see the paste-ups as one big mess. "That looks like hell," he told them. "There's no way those things will ever print. Just start over." Not being argumentative, the staffers obeyed instructions and threw away several thousand dollars worth of perfectly good paste-up sheets.

HASTY PURCHASING OF THE WRONG EQUIPMENT

There's an immense amount of impulse buying in the systems game. Owners of a design firm will hear some success stories from colleagues, maybe hear some glowing lecture at a professional convention, and get the idea in the back of their minds that they had better "keep up with the times."

Then a big job comes along and the boss thinks: "Great, with money like that we can get this office up to date." Shortly thereafter a sales representative comes along to offer a very high-priced piece of equipment. "This," says the rep, "will do everything there is to do in systems drafting." Actually it won't, but the boss has no technical knowledge with which to make that judgment. So they buy it. It takes a while to learn how to use the big machine. Eventually it dawns on everyone in the office that this equipment won't produce for them, and they stop using it.

The equipment sits, unused. Unfortunately there's a status problem involved. If anyone were to say out loud that the equipment doesn't work, the boss would lose face, and the informant would lose his or her job. Silence is golden.

As design firms computerize, errors are made. Then the errors are compounded instead of being corrected. The most common problem is in firms that have been slow to change to systems. They go computer to catch up quickly with the latest graphics and drafting technology. Unfortunately their knowledge of systems in general and computers in particular is, at best, primitive.

They buy the wrong equipment. It won't perform and the firm should start over. But to do so would require management to admit to error. So they don't. And at this point, since they have all that great computer drafting capability, nobody in the firm can be allowed to introduce the simpler reprographic systems that would augment the big machine or, on most jobs, outperform it. To do so would make management look bad. And that's not allowed, no matter what the cost to the firm.

5

TOOLS AND TECHNIQUES— THEY GET BETTER AND BETTER

THE VACUUM FRAME

An ingenious device has changed drafting room practices everywhere. It's the "flatbed printer," also called a "vacuum frame," "contact printer," or "exposure unit." It's simple enough—just a box with lights in it—but its impact is revolutionary.

The virtue is in "contact printing." The regular diazo machine (the ammonia-developed job print process that makes your blue line or black line prints) is a contact printer. But it's rotary. Print paper and original tracings are pressed together and exposed to light through rotating cylinders. Flatbed units, on the other hand, expose a whole drawing and print medium to light without moving them. The sheets rest on a glass plate or cushion that's lighted like a light box or light table—only much brighter.

HOW TO USE IT

By way of illustration, suppose you assembled a composite drawing of standard details, potentially a great time-saver. But the savings may be lost because of complications: It's a pain to do further regular drafting on the paste-up sheet, for example. It's treacherous making prints or a reproducible of a paste-up on a standard diazo

machine. And making a photographic reproduction is expensive, inconvenient, or both.

You need a way to make a draftable reproducible of the paste-up—one at exactly the same size and one that's cheap and convenient to make. Here's what to do:

1. To make a reproducible print of the original paste-up, place the original plus light-sensitive polyester on the copy bed of the flatbed printer. Close the lid. Switch on the vacuum pump so that the sheets will be pressed tightly together. Turn on the printer lights and timer (or light integrator).

2. After a minute or so of exposure time, remove the paste-up and the print sheet. Flatbed printers don't have their own ammonia developers, so you have to run the print through the developer part of a regular diazo machine to make the printed image visible.

3. Use the polyester reproducible for further drafting. For ease of drafting the reproducible should be erasable or moist-erasable polyester. Besides finish-up drafting, you'll use the reproducible to run final diazo job prints on the regular rotary diazo machine.

In this three-step example, you obtain a reproducible of a paste-up in-house within minutes. And you get a result that used to take a day—or days—to get through photoreproduction services.

41

Vacuum frames made for architectural and engineering systems graphics have problems. There are complaints about noise, voltage and timer irregularities, delivery and service delays. There's always some break-in and learning time before the devices start paying off. Still, they're comparatively low-cost for what they do—$2000 for a small tube-type unit up to $7500 for a rugged, high-powered top-lighted model (1982 dollars).

CHARACTERISTICS TO LOOK FOR IN A VACUUM FRAME

Good light spread. You may have to add some aluminum foil, extra lights, or diffusers to avoid hot spots and falloff around the edge of your prints. (Some presumed falloff, a darkening at the borders of some prints, has proved to be due to accumulated dirt and oil from extensive handling of the drawing sheets.)

Light power. It's ultraviolet light that is active in diazo printing, so don't assume that any old reprographics vacuum frame will do. Some use sodium vapor or arc lights, for example, which are plenty bright but no great shakes with diazo emulsion.

A light integrator. Some locales suffer serious variations in electric line voltage. A vacuum frame that has only a timer to control exposure times may be unreliable during fluctuations in line voltage which alter the print light being emitted. A light integrator measures the total quantity of actual light during an exposure process, not just the time, and exposure time is automatically modified to meet varying voltage conditions.

Availability of maintenance and service. This industry includes many newcomers who appear and disappear rather abruptly. Somebody has to be responsible for watching over your unit and making good when it doesn't.

Although the best models of vacuum frames are designed as heavy-duty production tools for the graphics and printing industry, they should not be used as production diazo printers. They cannot turn out your regular blue line job prints efficiently. For that, use your in-house rotary printer or repro shop services. The value of the vacuum frame is in making accurate *single*-copy check prints of paste-ups and overlays, base sheets, and finished reproducibles which are used as the "original" for doing fast job printing on high-volume rotary diazo printers.

If you want to experiment with what a vacuum frame will do, you can make your own "solar-powered" unit. Get a sheet of glass or heavy clear plastic and a rubber or plastic mat mounted on a piece of plywood. Haul the unit outdoors and make solar exposures of your originals onto print media. And if you don't happen to have an ammonia-developer diazo machine, get a jar of ammonia and a plastic tube, closed at one end, in which to insert exposed prints to develop. Mount the tube over the ammonia jar, hold your nose, and start developing.

ADVANCES IN APPLIQUÉ GRAPHICS

"Appliqué" means graphic tapes, pattern films, rub-on letters and symbols—things adhered to a drawing rather than drawn. It can be considered a branch of composite or paste-up drafting.

Large-scale use of appliqués used to be an impossible headache. Rub ons or tapes would come off in diazo print machines. Drafting tools would snag against tapes and stickyback sheets. Stickybacks left patchy ghost backgrounds on diazo job prints.

The old problems have for the most part disappeared, eliminated by new techniques, materials, and equipment. Here's what's being done, followed by problems and solutions:

- Tapes for complex line work. A single tape line is no faster than a drawn one (although it's much faster to pick up and move). The main speedup comes when doing complex multiple line work, such as walls on a plan that shows construction materials and fire wall ratings. Pattern tapes are also used for piping; ductwork; utility, fence, and property lines on site plans; cut lines, break lines, etc. Besides this factor of speed, drafters favor tape because of its ease in making changes, sharper and denser line work on prints, and perfect graphic consistency, regardless of who puts it down.

- "Ruby red litho tape" to show wall construction. The tape is opaque in printing but translucent to the eye. That makes it easy to apply a length of wall and then backtrack with a blade to cut openings for doors and windows. The tape is widely available and comparatively inexpensive. To answer the problem of showing different kinds of wall construction when using a single black printing tape, many firms now key wall construction with small bubbles and numbers which are referenced to a wall construction schedule or to wall construction details.

- A design firm that's gone all the way in paste-up and appliqué drafting found a good substitute for the razor blades and graphic cutting tools. Those tools wear out quickly, especially when cutting polyester elements. So this firm now buys scalpels from a medical supply company. They pay for themselves through long life and lack of cut damage that sometimes occurs when people use worn blades.

- Rub ons for conventional symbols. Architects and engineers endlessly draw and redraw a universe of small symbols. They're fast to draw, to be sure, but faster to rub on—5 to 10 times faster once a complete appliqué system is created.

The rub ons (dry transfer) pay off especially for engineers in depicting the numerous symbols of their profession: fixtures in ceiling plans, switches, outlets, junction boxes, for electrical plans; fans, diffusers, risers, dampers, etc., for HVAC; small-scale I-beam and WF sections, connectors and notes that appear over and over on structural drawings; fixtures, valves, drains, pipe size notes—the multitude of similar repeated symbols that pervade plumbing drawings.

RUB ONS AND STICKYBACKS

The small elements ("too small to worry about") eat up years of drafting time in any drafting room. Efficiency-minded offices use commercial rub-on products that include many commonly used symbols, and they design their own custom sheets for the rest. An alternative favored by many is die-cut stickyback sheets upon which have been printed a myriad of small commonly used symbols. "Die-cut" means that the symbols are already cut on the stickyback sheet. You just slip a blade under a symbol, peel it off, and stick it on. The thin acetate used for these products doesn't ghost, and the elements—light fixtures, valves, pumps, plumbing fixtures, etc.—are easier to apply than dry transfer and, unlike dry transfer, are repositionable.

Manufacturers produce custom dry-transfer sheets or the precut sheets of any design you want. Per-sheet prices become economical in larger printings, of course, so some firms team up with others to make shared bulk orders from appliqué companies.

Many firms now make their own custom dry-transfer sheets with processing kits sold at drafting and art supply houses. They can be expensive on a per-sheet basis, but offices say the speed and convenience of in-house processing makes up for it.

Look for applique company catalogs at your local drafting, graphics, and art supply stores for more information on the latest products. These catalogs are an education in themselves. There's a list of names and addresses of appliqué sources in *The Guidelines Reprographics and Computer Resources Guide* cited in the back of this book.

OLD PROBLEMS, NEW SOLUTIONS

A note on speed: Appliqués can be slow—especially at first. There are tricks; for example, unroll graphic tape by holding the free end and pull the roll away from you instead of the other way around. And use ink, not tiny tape pieces, to show small segments such as door jambs. Initial clumsiness with appliqués disappears in a week or less. Then the process accelerates to become clearly faster than hand drafting.

To avoid the conflict of drafting tools snagging on appliqué products or otherwise damaging them, don't mix the two operations. Do appliqué work on an original paste-up sheet, make a reproducible, and then do the line drawing separately on the reproducible or on an overlay. The overlay option is especially convenient for consulting engineers, and it augments other efficiencies of overlay drafting.

Three major problems with appliqués have one solution. The problems include those which occur with mixed drafting as mentioned above, plus appliqué failures in rotary diazo machines and ghost shadows resulting from stickyback films. The single solution is flatbed vacuum frame contact printing.

A flatbed vacuum frame has no hot moving parts that pop off appliqués or paste-ups. It just sits there like a high-powered light table. Besides solving pop off and tearing problems, the flatbed vacuum frame enables you to conveniently expose paste-up and appliqué original sheets to reproducible media for a longer than normal period of time. Original line work has to be dark and dense (as it is with ink and appliqués) for this to work. By exposing the originals for a longer than normal print time, you burn out ghost shadows both of stickybacks and of clear, taped-on films. The reproducibles—made on erasable diazo paper or film—come out like clean, haze-free original tracings.

The in-house flatbed printer isn't mandatory for extensive appliqué use—it just helps a lot. Architects and engineers work out alternative systems: appliqué and paste-up sheets (covered with clear polyester), gently shepherded through rotary diazo printers; use of repro shop vacuum frame printing or photo systems. All are available as options or as interim testing or transitional steps.

PHOTODRAFTING

KEYNOTING

Photos are extremely desirable when dealing with existing conditions such as site work, neighboring buildings, remodeling, adaptive re-use, retrofits, etc. They're desirable because they show what is really there and show it more clearly than any drawing can. They're desirable because they'll let you complete some documents 2, 3, even 10 times faster than traditional measuring and redrawing processes.

There's been resistance to photodrafting from those who are ideologically married to the drawing process. And there's been resistance from those who haven't appreciated the wide scope of usefulness of combining photos, drawings, and text in the same presentation or working-drawing package. As an example of this versatility, some firms now do their standard details in the form of partial photos. Photos are made of existing standard construction, walls, stairs, windows, parapets, etc. Then the photos are cut, and the insides of the construction are drawn in as an x-ray view of how the detailed item works and how it relates to other construction.

Photos are also great for showing exactly what is wanted in such hard-to-specify items as decorative wood grain, stone masonry, special concrete textures, custom tile work, and wood joinery. Useful samples of finishes and construction await on every block in every city for anyone willing to take their pictures.

Keynoting is one of those multibenefit techniques that just don't quit. I described the basic processes and efficiencies of keynoting in *Systems Drafting*, and they are described elsewhere in this text. Before I describe the advances, here's a review of the basics:

Keynotes constitute a legend of numbered typewritten notes placed at the right side of a drawing sheet. The numbers with arrows are inserted in the field of the drawing where the notes would normally be. Keynotes list just once a note that might otherwise appear dozens of times on a drawing. They're typewritten instead of hand-lettered, which improves the time savings even more. As a file is built up, the notes become standardized and reusable like standard details. They're not suitable for all types of drawings, but they work especially well for exterior elevations, sections, roof plans, photo drawings, and many consultants' drawings. The guiding rule is to use them on drawings where a relatively few notes are repeated many times.

Two things are happening to augment keynoting time savings even more: (1) offices are storing keynotes (as well as assembly notes and specifications) in their computers, text editors, and word processors, and (2) architects and engineers are numbering their keynotes to correspond with specification section numbers. For example, if one is using the Construction Specification Institute (CSI) specs format, masonry keynotes are numbered under a 4.00 "Masonry" subheading on the keynoting column, and then listed with a "4" prefix: 4.01 4.02, etc. Notes pertaining to "Finishes" are listed 9.01, 9.02, etc. After a file of keynotes is created, drafters select notes from the master list and jot down keynote numbers as needed, instead of writing out complete notes. Typists later complete the keynote lists by matching numbers to the notes in the master list.

As a further improvement to keynoting, a civil engineer and landscape architect use keynotes on their site plans and, where appropriate, precede each keynote with a related detail key bubble number. Ultimately, details, specification sections, and notation are linked by the same identification numbers. Contractors respond very well to this kind of documentation.

One wrote to congratulate an architect for using integrated notation format, saying it was "extremely helpful to us and the subcontractors in identifying and categorizing the components of the building."

I have carried the keynote process further, as you'll see in the last part of this book, and integrated it with detail numbers and other data as well as specifications. More on that later.

UPDATE ON INK DRAFTING

Why do architects and engineers who draw with ink say it's better and faster than plain old drafting lead? Many readers will remember the old ink ruling pens; they'd splotch at the twitch of an eyelash. Then the improved technical pens came along. They were fine for special illustrations, but users had to spend half their time at the washbasin cleaning and unclogging them. Now it's all turned around.

OVERCOMING PROBLEMS AND CONCERNS

Inking is as fast or faster than graphite lead drafting, despite the need (at first) for light pencil guidance layouts. The need for pencil guidelines rapidly diminishes with practice. Time studies at industrial and engineering drafting rooms show that lead and ink drafting run neck and neck. Many architects and engineers have told me inking is actually faster than pencil lead once people adjust to it.

The new inks dry fast—almost immediately. Drafters usually tape coins or thin pads to their parallel bars or drafting machines to lift the tools above the drawing surface to guard against smearing wet ink. Many find they don't need that much protection because the inks dry so fast. It is useful to use triangles with small bosses or "bumps" on them, however, to keep the triangles above the drawing.

One reason ink drafting is ultimately as fast or faster than graphite is that it erases so easily. Ink allows easy 100 percent ghost-free erasing. It's far cleaner than erasing graphite and much faster. There's a catch. You get the easy erasing by drawing on polyester (Mylar) drafting media—instead of on tracing paper.

Inking on polyester is a matter of laying a film of ink across the surface of the plastic instead of engraving it into paper. That's another reason for the comparative speed—less effort. And it's why the erasing is fast and ghost-free. You don't have to scrub graphite or ink particles out from between the tracing paper fibers. Inking allows you to gain the advantages of polyester without the pains. More on that in a moment. Some firms avoid using polyester because of "cost." The difference between 1000H tracing paper and comparable polyester drafting media is a couple of dollars per 30-inch by 42-inch sheet. The cost difference is offset several dozen times because of durability, rapid printability, easy overlay viewing, and other qualities of the plastic drafting film.

Special erasers such as the Pelikan PT-20 make it all even easier because they contain an erasing fluid imbibed within the eraser. They erase exceptionally well and quickly.

It takes time to adjust to inking—mainly because of people's fear of making errors. As I said in *Systems Drafting*, the timidness and shakiness leads to errors which reinforces the fear. After a while the inker realizes that mistakes don't matter and are less trouble than if made with graphite. It takes one to two weeks to bring drafting speed up to normal, and then it starts to go faster than what used to be normal.

Hand-lettering is hard to do with the technical pens, so some users switch to a compatible alternative medium when doing notation. One alternative is to use a strong black, moist-erasable, felt-tip pen such as the Sanford Vis-a-Vis. It's easy to letter with, and although the image isn't quite as opaque as the ink line work, the comparative darkness evens out when a whole inked sheet is converted to a sepia line reproducible. Another alternative for hand-lettering preferred by some drafters is the plastic lead that's designed for better adhering quality to polyester drafting media. Keep in mind that plastic lead in a wood pencil doesn't wiggle as much and isn't as likely to snap as the brittle leads in mechanical pencil–type lead holders.

ADVANTAGES OF USING INK

Ink works if you switch to polyester media at the same time. Offices that have wanted to switch have sometimes been hamstrung by problems with special leads and erasers. There are complaints about the unpleasant "feel" of drafting on plastic, too. Ink on plastic, on the other hand, has a good feel. It's a fine, controlled glide—fast and free of strain.

There's no spreading of a graphite haze over a tracing. Drawings remain pristine—very fast to print on the diazo machine. And they're excel-

lent for photoreproduction. Lines aren't engraved or repeat-stroked (which is one reason why ink drawing goes fast).

A major advantage is that ink drawing on polyester goes hand in glove with the new reprographic systems, such as overlay and composite drafting. Ink drafting is much more readable and clear in contact printing or photoreproduction than is graphite drawing. It helps expedite conversion to new drafting systems.

Finally, a couple of pointers to make sure the pens work:

1. Draw on polyester, as we say, and use nothing less than the jewel or carbon-tungsten tips. Regular steel tips will wear out in no time on polyester.

2. Use the new pen tip protectors that maintain clog-free performance, and purchase an office ultrasonic cleaner ($70 to $120) for superfast cleaning, if and when the pens do clog.

Fig. 5-1. Dry transfer, or "rub on," used for small, often
used electrical and plumbing symbols.

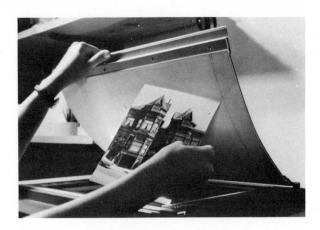

Fig. 5-2. This is a low-cost photodrafting process that you can handle with common office equipment. It starts with a photo printed on standard opaque photo print paper. The photo is first copied with a regular plain-paper or "bond" office copier that uses dry powder toner.

Fig. 5-3. Plain-paper copiers will copy onto tracing paper or vellum as well as onto regular bond copy paper. Here a 16-weight sheet of ordinary tracing paper is inserted in the paper feed of the copier.

Fig. 5-4. The image from the original opaque photograph has now been copied onto translucent tracing paper. This tracing paper print can now be run through a standard diazo print machine.

Fig. 5-5. The tracing paper copy of the photo is run through the diazo machine and copied onto 3-mil sepia line polyester print film. The reason is that the toner image fused onto the paper is not, in itself, strong enough to reproduce well on a paste-up carrying sheet. The image has to be intensified by converting it to the sepia-emulsion image.

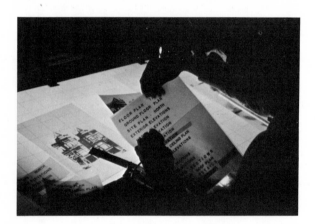

Fig. 5-6. The sepia line polyester film copy of the photocopy is resting on the light table. The graphic work assistant is selecting titles and other appliqué drafting products and tools to be used to make this photo drawing.

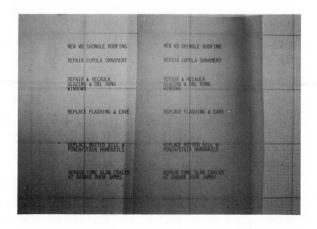

Fig. 5-7. Notes for the photo drawing are first typed on tracing paper with a regular electric correcting typewriter using film ribbon. These notes so typed are not themselves dark or intense enough to use in a paste-up, so the typing is intensified by running it through the diazo machine onto sepia line film, just as was done for the photo. (Orange carbon paper used to be applied on the back side of typed tracings for the same purpose—to intensify the typed image for diazo printing.)

Fig. 5-8. The photo drawing is split down the middle, the notation block added, and translucent dotted tape used as notation leader lines.

Fig. 5-9. The paste-up photo drawing can be reproduced again onto sepia line polyester film either with a slow speed run through the diazo machine or on a vacuum frame. Since it's on draftable, erasable film, any further changes can be made directly on the reproducible. The paste-up can be assembled on a larger paste-up sheet or printed in 8½-inch by 11-inch format for a project manual.

Fig. 5-10. When printing a paste-up like this, it's best for the emulsion side of the paste-up elements to be set against the emulsion side of the reproducible sepia print film. That way the copy will come out right-reading with the draftable surface on top, erasable emulsion on the back.

51

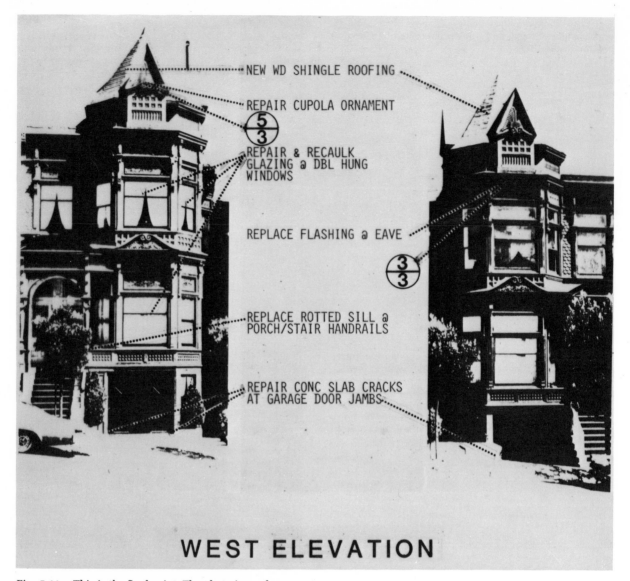

NEW WD SHINGLE ROOFING

REPAIR CUPOLA ORNAMENT

5/3

REPAIR & RECAULK GLAZING a DBL HUNG WINDOWS

REPLACE FLASHING a EAVE

3/3

REPLACE ROTTED SILL a PORCH/STAIR HANDRAILS

REPAIR CONC SLAB CRACKS AT GARAGE DOOR JAMBS

WEST ELEVATION

Fig. 5-11. This is the final print. The photo image has gone through photocopy and two stages of diazo printing. Even after three generations, it's still readable and adequate for diazo job print working drawings.

If you wanted to reduce the strong contrast, you could use a "copy screen"—a transparent plastic sheet with white dots printed on it like white zipatone—when copying the original photo on the office copier. That breaks the photo image into a sort of "poor person's" halftone which subdues the black areas of the photo and makes them print grey.

Another option is to have photos processed on the thinnest photo paper available. It's thin enough so that you can get a direct contact print onto sepia diazo polyester without going through the office copier stage of converting an opaque original into a translucent copy for diazo printing.

Fig. 5-12. Typewritten notation doesn't require any but normal office equipment when you use translucent paste-ups. Here, a list of keynotes is typed on regular tracing paper with an electric correctable typewriter using a film ribbon.

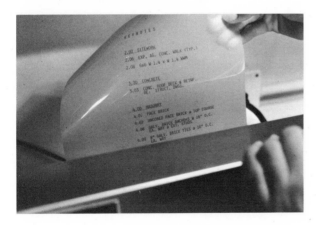

Fig. 5-13. To intensify the image of the typing for good reproducibility, it's printed onto the sepia line diazo polyester film. This "second original" will print better than the original because the letter images are more opaque to light than those of the original are.

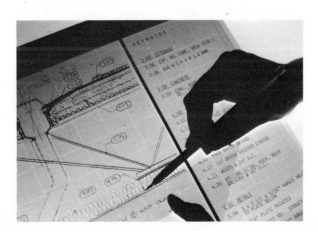

Fig. 5-14. The keynote list is spliced into the working drawing paste-up sheet. Some space is provided if it's necessary to letter-in notes by hand later. The notes are clustered and numbered according to the CSI specification divisions.

PART 2
ADVANCED OVERLAY REPROGRAPHICS

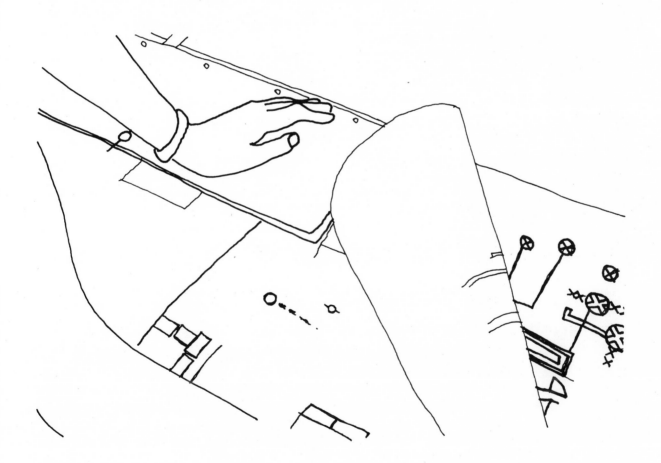

Illustration by Lanée La 'shagway.

6

OVERLAY DRAFTING, NICE AND EASY

ESSENTIALS OF OVERLAY

Everyone has done overlay drafting whether they know it or not. Any time you have laid paper over an existing drawing, either for tracing or to pursue a design process, you have done overlay. Contemporary pin register overlay drafting just elaborates on that process, formalizes it, and at the same time generates some remarkable graphic and coordination benefits.

The benefits include:

1. The opportunity to test out more solutions to plan and design problems and to show more options in presentation drawings, without spending extra time on the process.

2. A big reduction in repetitive drafting even for small-building projects.

3. Much improved consistency between drawings produced by architects and those done by consulting engineers.

4. The chance to use special printing techniques that sharply differentiate and clarify different kinds of information on the drawings. For example, you can print the walls of your floor plan as a subdued, "screened," background-dotted image while the feature data of that drawing appears in contrasting bold solid line. Or, you can unmistakably differentiate two or more kinds of data by printing drawings in different color inks with offset printing.

That's an important array of potential benefits. But they don't come without some problems. You may have heard about, or experienced, some of the overlay drafting problems: confusing numbers and combinations of base and overlay sheets, complex and costly reproduction processes, resistance by consulting engineers and their staffs. Without dealing with these issues in detail right now, let's just say that all these problems had specific causes and they've all been taken care of.

The essential components of the overlay drafting process are the base sheet and the overlay sheet. The base sheet is a print, on polyester film, of repetitive data. For example, a floor plan, used as background for a number of different engineering drawings, is repetitive. Instead of having different drafters redraw the plan as part of their work, the plan is turned into base sheets by making multiple copies onto clear sepia line "slick" films. Drafters get ready-made copies instead of having to redraw them.

The overlay sheet is an original drawing. The base sheet laid on the drafting board is a copy, acting as what we call a "constant" of reference data. The overlay will contain the unique, or variable information, such as electrical, plumbing, and HVAC systems.

Both base and overlay sheets have to remain consistently stable in size relative to one another. To assure stability, we use polyester (Mylar) for both the printed base sheets and the drafting overlays. Polyester will not stretch or shrink as tracing or print paper does.

Since the image drawn on the overlay sheet has to remain stable relative to the base sheet, both sheets have to be kept in precise registration with each other during drafting and printing. For that we use prepunched holes across the top of each sheet and pin bars that match the punched holes. The pin bar keeps base and overlay in registration in a very direct physical way, independent of visual alignment and registration.

The one-on-one base sheet and overlay sheet relationship is fairly straightforward. So where's the complexity people complain about? Mainly in the job planning and printing processes. Although you only deal with one base sheet copy and one original overlay in the drafting process, you may have to combine three, four, or more different sheets when making a single print. For example, a final print of electrical work may involve the floor plan, a separate "format sheet" with title block and border, the reflected ceiling plan as an intermediate overlay sheet, and an electric power drawing as the final overlay in the print sandwich. Multiply that multisheet complexity in printing by several dozen or several hundred times and you can see the potential for mix-ups, time lags, lost sheets, and high repro costs.

We're now going to review several uses and phases of overlay work, from the simple to the complex. We will lay out the fundamental steps of overlay, one by one, and in the process name the major "dos and don'ts".

THE FIRST STEP, OVERLAY IN DESIGN DRAWINGS

The key to efficient overlay drafting management is in the concept of "stop points." For example, if you're doing rendered presentation drawings of plans and elevations, you stop work on the drawings before doing any rendering. Stop work and then proceed with overlays for all the rendered titles, shades, shadows, furniture, etc. That way, when you later proceed with the project, you can reuse the uncluttered original base sheet directly in working drawings.

The question always arises: Why not just make copies of the plans and elevations as backgrounds, and then do your renderings directly on those? In some cases that will work fine. But often there will be some changes in the original work, or a need to show options and variations. When you make changes, you will have to revise both the rendered drawings and the original—double the work. If you show options and variations, you will have to make even more changes on even more drawings every time there's a change in the original concept. When using the base-overlay system, you reduce the total number of hand-drawn changes that have to be made in any revision situation. Revisions are made once, on an original base, and further copies are made through printing processes instead of hand drawing. That eliminates the errors and discontinuities that are likely to creep in when people copy by hand.

THE NEXT STEP, STOP POINTS FOR CONSULTANTS' BASE SHEETS

After design development and presentation drawings are completed, architectural floor plans are firmed up for use by consultants. That means the floor plans are completed only up to a limited point, where they contain only information of common use by both architect and engineers. That means the plans do not yet have strictly architectural notation, door symbols, window symbols, finish schedule symbols, interior dimensions, etc. All that architectural data is reserved for the architectural overlays.

Plans are brought to a partial state of completion without the strictly architectural data and then copied for use as base sheets by consultants and architectural drafters. This stop point of completion excludes data not important to the engineers and includes data that is. For example, the plans show door swings because that's important to electrical consultants for coordinating their light switches. But they don't include door schedule reference symbols because that information is not needed by the consultant.

The floor plans will include rest room fixtures, drinking fountains, hose bibbs, slop sinks, floor drains, fire hose cabinets, plumbing chases and access, structural elements, etc., for the plumbing consultants' drafters. Reflected ceiling plans will show sprinkler heads when set by the architect for design reasons.

For the mechanical engineers, plans will include shafts, equipment to be vented, door swings, mechanical equipment, mechanical equipment bases and curbs (built-in and not part of the mechanical contract), changes in floor and ceiling levels, and structural elements. Reflected ceiling plans will show changes in ceiling levels and outlines of mechanical fixtures and equipment.

Besides door swings, plans for electrical consultants will show locations of electric closets and panels, shafts and chases for electric and communications wiring risers, electrically powered equipment (such as clocks, automatic doors, fans, etc.), special light fixtures (such as for fire exits), concealed lighting, fixtures, switches, and outlets that are specifically located by the architect for design or planning reasons. Reflected ceiling plans show fixtures, access panels, hanging electric fixtures and equipment, etc.

Structural work is a special case. All the other engineering disciplines require the same basic data as that listed above, but the structural work precedes and underlies most of the other items. So at first structural drafters get very rudimentary base sheet data: mainly basic building dimensions, foundation outline, and structural grid. Base and overlay coordination comes into play a little later as architect and structural engineer coordinate locations and sizes of slab openings and depressions, raised floors, curbs, stairs, elevator shafts, drains, etc.

BEYOND THE STOP POINTS—BASE AND OVERLAY COORDINATION

Just as the consultants' drafting staffs receive copies of the original floor plan brought up to the stopping points just described, so do the architectural drafting staff. They continue adding the unique architectural data on their overlays: finish schedule keys, door and window schedule keys, detail and section bubbles, interior and exterior dimensions, fire walls, material indications, cabinets, built-in furniture, wall-mounted panels, shelves, etc.

The question often arises as to whether room names should be on the base sheet or on an intermediate overlay. The argument for the overlay separation is that room names might overlap and conflict with consultants' drafting. In general, we favor including room names on the base sheets. It is "constant" data of use to consultants, and since all base data will be screened and subdued in final printing, it won't matter if there are overlaps. Hard line data by the consultants will stand out clearly even if drawn or lettered over base sheet notes or titles. For that reason it's not a major problem if architectural drafters add more information than is strictly needed for the engineers' base sheet stop points. It's very desirable to hold to the stop points, of course, but you do have some hedge factor because of screening if someone goes a little too far in original base sheet drafting.

Inevitably, as working drawings proceed, there will be revisions in the architectural plans that affect the work of the consultants. Most changes will be minor and won't require a new submittal of revised base sheets. Information about changes is sent to all drafters via memo and marked check prints, or by prints of revised portions of the plan to be used as slip sheets under the base sheets. The revised area of building is slipped beneath base sheets, and if that affects the drafter's work, he or she can make the necessary changes. If it doesn't affect the work, then nothing more needs to be done.

After a while enough changes may accumulate to make it necessary to send out new base sheets to everyone. At this point it's mandatory that all currently used base sheets be returned, logged in as returned, and destroyed. If there's no record and control of base sheets, it's inevitable that drafters will accidentally use a rescinded one and be working over obsolete information. Although the coding and tracking of base and overlay sheets are management functions, everyone on a project should be responsible for guarding against such hazards as obsolete base sheets.

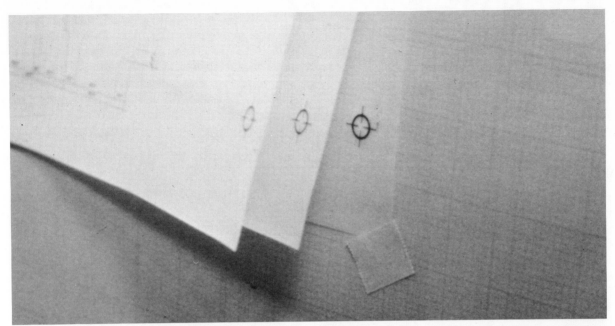

Fig. 6-1. This illustrates the principle of registration in graphics. "Targets," "bull's-eyes," or "registration marks" are drawn or stuck onto the base sheet. All subsequent sheets drawn in registration with the base sheet have registration marks added exactly on top of the original marks. Henceforth, you can always check visually to see if base and overlays are in registration or not. At least three marks on each sheet are required to ensure accurate alignment.

If you're starting on a small job and just want to test out the overlay concept in principle without being bothered with pin bars, you can use registration marks instead.

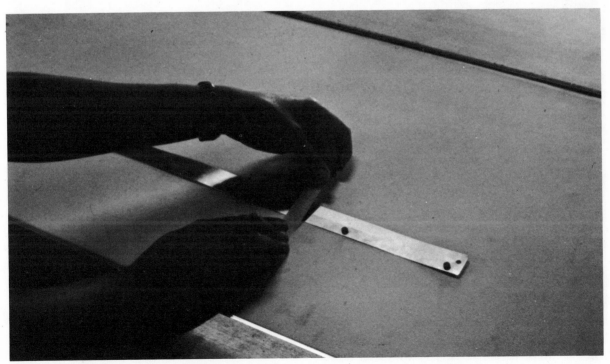

Fig. 6-2. To prepare the drafting station for pin register overlay drafting, align the pin bar with the parallel bar or drafting machine bar and tape it into position.

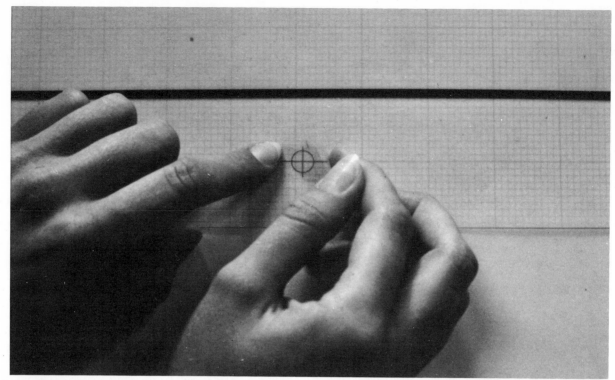

Fig. 6-3. Add registration marks at the bottom of the drafting work area. That's a reminder to put a couple of registration marks at the bottom edge of base and overlay sheets. Although the pin bar is primarily responsible for registration, these marks provide a visual double check. And they provide a check after sheets have been printed.

Fig. 6-4. When it is time to start an overlay sheet, first lay down an up-to-date copy of the base sheet. It's usually in the form of a clear, non-matte sepia diazo polyester called a "slick," or "throwaway."

Fig. 6-5. This series of photos shows how base sheets are made. First the original base sheet is laid in registration with sepia diazo "slick" print sheets.

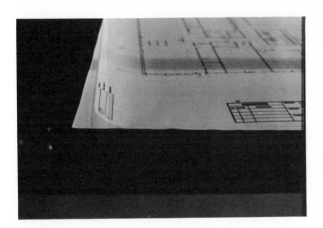

Fig. 6-6. To get exact size fidelity, press the original and print media together in the vacuum frame thus making a direct contact print.

Fig. 6-7. To develop the base sheet copy, use the ammonia developer of the diazo machine. The same number of base sheets are made as there will be drafters working on floor plans. Base sheets go to architectural staff members so that they can continue with detail keys, door and window symbols, architectural notation, etc. Base sheets go to engineering staff so that they can add their specialized overlay data. A log must be kept to record who has which base sheets, when they're updated and new copies issued, etc.

Fig. 6-8. To make a combination print of two or more layers, lay the most important or dominant layer of information directly on the print sheet. Check that the "print control sticker" shows the right number and sequence of layers.

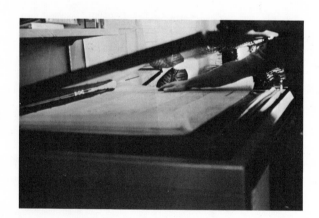

Fig. 6-9. A powerful vacuum frame like this can make a composited print of four to five layers. The total number depends on the translucency and thickness of the overlays, the original line quality, and various attributes of the vacuum frame light system. If you have more layers to combine for one print than can be made at one time, gang them by combining, say, three onto one reproducible, another three onto a second reproducible; then combine the two reproducibles.

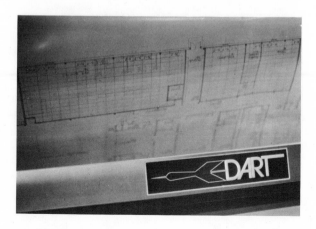

Fig. 6-10. The layers shown here have been exposed on the vacuum frame and now emerge as a single image either on diazo print paper if you're doing a check print, or on sepia line reproducible paper or polyester.

Fig. 6-11. This shows the sequence of how changes made on the original base sheet are conveyed to all those who are working on overlays atop reproductions of that base sheet. First some door locations have to be revised. (Note that although an electric eraser is not recommended for use on polyester, it is still commonly used. Revision requires a plastic eraser, a light touch, and some moisture on the eraser to prevent the heat of friction from damaging the matte drafting surface.)

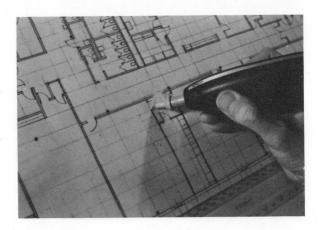

Fig. 6-12. Diazo paper contact prints are made of the revised portion of the base sheet floor plan. These will be distributed to all drafters working on overlays connected with this floor plan.

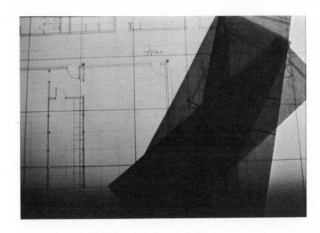

Fig. 6-13. Along with slip sheets showing the revision goes a red-marked reference print to show the items that have been revised in the context of the total plan.

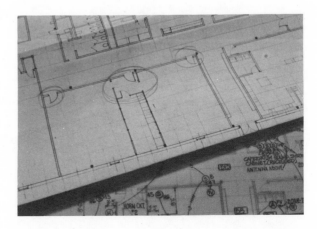

Fig. 6-14. The drafters lay the slip sheets into position under their base sheets. Now they can read the change through the base and overlay sheets.

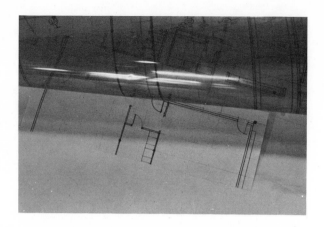

Fig. 6-15. The modified door locations don't affect the plumbing consultant's drafters. They are made aware of the change, but they don't have to do anything to their work to reflect the revision. The electrical consultant's work, on the other hand, is affected. TV outlets that had been on the walls now have to be moved to allow for the new door locations.

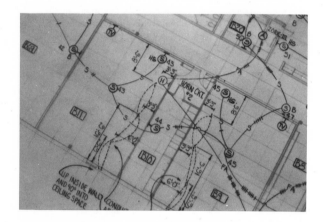

Fig. 6-16. The electrical consultant's drafter revises the affected area of work but doesn't have to worry about erasing or redrawing any of the architectural modification.

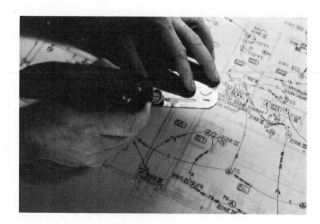

Base and Overlay Guidelines

Basic schemes

Civil: The Civil Department chooses to use:

1. A preprinted border sheet of line value (100%)

Plus: 2. An architectural or structural base drawing of screen value (30% or 50%) (1)

Or: 2a. A survey base drawing of screen value (30% or 50%) (1)

Plus: 3. A civil overlay including dimensions, objects, notes, etc., of line value (100%)

Equals ⟶ **One civil drawing**

Architectural: The Architectural Department chooses to use:

1. An architectural base drawing consisting of structural grids, walls, windows, doors, room names and numbers, and other permanent objects, of line value (100%) (2)

Plus: 2. A preprinted border sheet with notes, dimensions, references, etc., of line value (100% (2)

Equals ⟶ **One architectural drawing** (3)

Structural: The Structural Department chooses to use:

1. An architectural base drawing for tracing purposes only (a throwaway), of screen value (30% or 50%)

2. A structural drawing drawn in its entirety on a preprinted border sheet, of line value (100%) (4)

Mechanical: The Mechanical Department chooses to use:

1. An architectural or structural base drawing, of screen value (30% or 50%) (1), (4), (5)

Plus: 2. A mechanical overlay on a preprinted border consisting of objects, dimensions, notes, references, etc., of line value (100%) (2), (5)

Equals ⟶ **One mechanical drawing**

Electrical: The Electrical Department chooses to use:

1. An architectural or structural base drawing; of screen value (30% or 50%) (1)

Plus: 2. An electrical overlay on a preprinted border consisting of objects, dimensions, notes, references, etc., of line value (100%) (2)

Equals ⟶ **One electrical drawing**

All other drawings in all disciplines

1. A preprinted border sheet consisting of details, sections, elevations, schedules diagrams, etc., of line value

Equals ⟶ **One drawing**

Footnotes

(1) All elements to be of screened value will be done by the print shop at the vacuum frame or the camera when the compositing is done, before the drawings are printed, from line value originals.

(2) Client criteria may dictate that Architectural will put room names and numbers on their last overlay instead of the base drawing. This will permit Mechanical and Electrical to place the room names and numbers on their last overlays in locations convenient for their respective purposes.

(3) Architectural reflected ceiling plans will be drawn on preprinted border. No screened value base drawing will be an integral part of that drawing.

(4) Structural will provide Mechanical and Electrical with a sepia diazo mylar or SDM (draftable and erasable), of the foundation plan object lines and structural grid upon reaching an appropriate stopping point. Structural will have the preprinted border reprographically blocked out from the SDM.

(5) Mechanical and Electrical Department will update the SDM provided to them by Structural for foundation changes after the initial issue.

Fig. 6-17. These illustrations in the John Graham & Co. in-house newsletter, *Shop Talk*, help clarify the overlay system to staff and encourage use of the system. This is an unusually clear presentation of how it all goes together.

(Courtesy of Jim McGrath, John Graham & Co.)

7

OVERLAY PROBLEMS AND BENEFITS

OVERLAY COMPLAINTS

Pin register overlay drafting has raised as many complaints as compliments among its users. The main problems, and their solutions, are:

The main complaint is that reproduction costs go sky high. Architects and engineers complain that some repro shops know as little as they themselves know about overlay reproduction and use the most expensive photographic repro methods to get results achievable at much lower cost through diazo printing. Photo blowback is expensive: $30 to $100 per sheet depending on whom you deal with. However, one-to-one contact printing on diazo products can give completely satisfactory results for $5 to $10 per sheet.

A second complaint is that there are too many complications. The source of these complications is excessive separation of overlays. The consensus now is to keep to one base and one overlay, where possible, to make one final drawing. Early users of overlay were separating plans into six, seven, and many more overlays.

A general problem in using overlays has been the resistance and confusion among drafting staff and consultants. Added to this have been assorted misunderstandings, too much downtime for printings, etc. There is a two-part cause for these problems: (1) lack of a detailed plan and mockup of the job process, and (2) failure to assign someone to monitor the overlay process and attend to its special demands. Those who should monitor haven't had the time to perceive this extra dimension of job management.

GAINS IN COORDINATION AND GRAPHIC CLARITY

Despite problems, confirmed overlay users will never go back to the old ways. There are many reasons why. First, overlay is a time-cost saver— if and only if you avoid excessive overlay separations and curtail expensive camera work. When it's done right, it's a big aid to coordination among the architect, consultants, and building trades. You can overlay transparencies and see conflicts and interferences that normally wouldn't show up until construction. Another value is that firms go to screened prints, offset printing, and color printing to improve the clarity and quality of their finish documents. Those require overlay separations. (Note that screened prints with offset printing require camera work—but at the end of a job and not for base sheet or interim printings.)

Screened prints ("shadow prints") get a lot of mileage. Screening breaks up the image of a drawing into small dots or lines, so that line work fades as background reference. "50 percent" screening is most commonly used. Data you want to emphasize are printed in solid line for contrast. A valuable checklist of special uses of screening to clarify final prints appears on page 70.

Drawing type	Screened background	Solid line for contrast
Renovation, rehab	Existing construction	Demolition, new work, additions
Foundation plan	Floor plan above foundation	Foundation, basement, crawl space
Roof plan	Floor plan below roof	Roof surface, drains, vents, etc.
Consultants' drawings	Architectural floor plan background	Separate building trades: electrical, plumbing, structural, etc.
Consultants' reference	Architectural background with furnishings, equipment	Electrical, communications
Plans, elevations, sections	Base construction	Alternative bid construction or additions
Site plan	Surveyor's map, existing contours, foliage, etc.	New contours, demolition, construction, site work
Consultants' site plans	Contours, new construction	Site electrical, mechanical, irrigation, landscaping details
Punch list reference	Construction as designed	Work requiring completion or correction
Change orders	Construction as designed	Revisions with cloud bubble or with revision symbol tags keyed to a revision schedule
As-builts	Construction as designed	Construction as changed

Inventive architects and engineers come up with numerous other variations, such as screened backgrounds for shop drawings, screened wall sections showing basic wall construction with detail variations in solid line, screened backgrounds for the client to facilitate drawing of future changes in the building, screened architectural backgrounds with special data required by regulatory agencies in hard line.

You can do your own shadow print, screened background drawings, or base sheets in house by diazo printing. Get hold of a 50 percent 120-line "tint" from your repro house (they can make contact prints in different sheet sizes from their originals). Or buy a tint screen from a photo repro supply company. Or, in a real pinch, splice together some large zipatone pattern sheets and make a contact print of those on sepia polyester as your homemade "poor person" tint screen. Then follow the steps shown and described in the illustrations that show two methods of in-house screening of base or background sheets.

Systems-oriented offices come up with a dizzying variety of ideas and new uses for overlay. Here's an update on some items originally reported in my book, *Systems Drafting:*

- Overlay drafting simplifies foreign language translations for overseas work. In one case, original base sheets were drawn without notation; overlays were first done in English, a second set of overlays was translated into Korean for the contractor, and a third set was done in Arabic for the client. Variable notes and dimensions on overlays can be photo-combined or separated from base sheets to suit any informational need.

- A new concept has emerged: "living documents" for clients. For some clients there's more drawing after construction than before, starting with "as-builts." Some buildings are subject to continuous changes after construction: new offices, laboratory alterations, etc., with all the attendant changes in electric and mechanical systems. Living documents provide an ongoing record of all changes. As-builts and "living documents" are reportedly far easier to do with overlay and composite drawing systems.

- A New York City office has gained numerous unexpected benefits. They've cut back on job-site interferences, subcontractor conflicts, and union jurisdiction problems by using a new kind of "coordination drawings." These are special, multicolor overlay composites—not part of bidding drawings—used just to show various disciplines and subcontractors how everyone relates to others' work.

- Further after-the-job uses include real estate and tenant brochures, publicity drawings, building maps and signs, office brochure drawings, and entries for journal publications and design competitions. Normally most of such drawing has to be done separately, started from scratch. But the reprographic systems allow separations, resizing, relocation, and re-use of all portions of design development drawings and production drawings for these other applications.

- "Systems" offices are gaining marketing advantages. They show their working drawings as if they were brochures. When showing the new graphics to prospective clients, they cite tighter bids, fewer construction claims, reduced errors, better jobsite coordination, etc., as selling points. While showing drawings that are generally done faster with systems, they also point to time and cost savings that accrue to the client because of the high percentage of reusability of drawings for "living documents," publicity, etc.

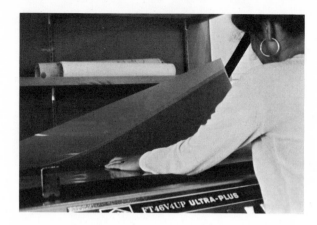

Fig. 7-1. To do an in-house screened shadow print background sheet, start first with a tint screen and the print sheet. This screen was "homemade" from Zipatone.

Expose the print medium to the tint screen on the vacuum frame or in a diazo print machine, but don't develop it yet. In so doing, you'll have a print sheet that is now broken up into a field of dots of yellow emulsion. In other words, you'll have a latent, undeveloped print of the tint screen.

Fig. 7-2. Now combine the still undeveloped print sheet with the drawing you want to print as a shadow print and make another exposure on the vacuum frame or in the diazo machine.

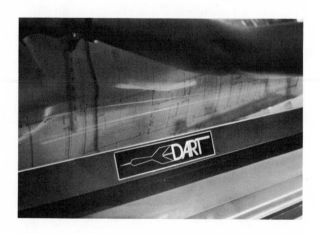

Fig. 7-3. Develop the print sheet in the ammonia developer of the diazo print machine and you'll have a screened shadow print. If you want, you can combine that screened image with an overlay sheet and make a new print showing a combination of screened base sheet and solid-line overlay.

Some people prefer to expose the background sheet image first and then combine that with the screen tint. That may work better in some circumstances. Also, you may have to practice a bit and modify the timing of the second exposure to avoid burning off too much emulsion from the print sheet and getting too weak a shadow print image.

Fig. 7-4. Here's a way to get a screened background or base sheet image and solid-line overlay on the same print sheet in one general operation.

First make a combination sandwich of print sheet, tint screen, *and* the overlay sheet you want to print in solid line. Make an exposure on the vacuum frame but don't develop it.

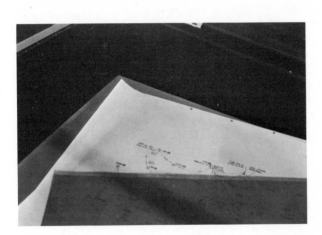

Fig. 7-5. The overlay sheet is attached directly to the print sheet because it'll remain in place after the first exposure and the tint screen on top will be removed and replaced by the base sheet.

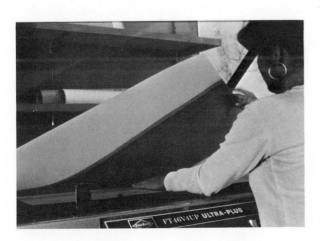

Fig. 7-6. Remove the tint screen. Leave the overlay in place on the print media. What you have now is a print sheet with both the latent image of the tint screen and a latent image of the overlay sheet you want printed in solid line. By remaining in place, the overlay sheet will cover it's own line work when the print sheet is reexposed to light to print the screened background image. When developed, you'll have a combination image of solid-line overlay and screened base sheet on one print.

8

COORDINATING THE CONSULTANTS

OVERLAYS

Traditionally, one of the worst problems in all but the simplest projects has been lapses in coordination between architect and consultants and between the work of the various consultants.

A starting point for the problems has been the custom of submitting prints of architectural plans to consultants so their drafters could redraw that work as background data for their own drafting. The redrawing process creates a built-in time lag right from the beginning. And it leads to errors as some engineering drafters misread or misdraw the architectural data. The errors are compounded when later architectural changes are not clearly transmitted to, or understood by, the consultants' drafters.

The traditional checking and coordination process adds another built-in trouble spot. The normal checking procedure for coordinating architectural and engineering drawings entails looking back and forth between check prints provided by all the job participants. The architect-coordinator sets up check prints from two, three, or more sources and visually compares them one by one. It's virtually impossible to avoid lapses and oversights in that kind of multiple, side-by-side comparison process. It hasn't been unusual for errors and omissions in coordination to slip through several stages of checking, through bidding, and all the way through to the jobsite.

Systems drafting, particularly the overlay system, eliminates most of the situations that lead to coordination errors. For example, as described in the preceding chapter, individual consultant drafters no longer have to redraw the architectural work. Instead they receive copies—mainly floor plans—in the form of sepia line polyester transparencies (slicks, throwaways) to be used as base sheets. No one redraws—or misdraws—the original data. Everybody on the job is working from exactly the same information. At this point in the work the coordination between and among all participants is 100 percent.

Since all overlay drawings of electrical, communications, plumbing, HVAC, and specialized architectural work have to be drawn in identical formats, scales, and positions on the drawing sheets, all drawings can be easily and precisely overlaid for coordination checking.

Instead of leafing back and forth for side-by-side, visual comparison of different prints, the coordinator can lay the overlays directly atop one another on a light table for direct, one-on-one comparison. Or the overlays can be printed together as special composited check prints. Either way, in layered form there's no way to overlook conflicts in construction. It's easy to spot drainpipes penetrating beams, plumbing through ducts, risers inside of rooms instead of inside chases, etc. If a consultant hasn't picked up a revision in the floor plan, that becomes obvious by layering overlays with the original base sheet.

Through layered comparison and printing it's easier to separate the work of different trades and contracts and still prevent conflicts. For example, preliminary communications and wiring plans have often been based on furniture plans which were based on preliminary floor plans which had long since been revised. Revisions don't make their way back through the cycle sometimes because it's too much trouble to redraw not only the revised construction but all the other work that hasn't been revised. Traditional drafting required that kind of total redrawing just to accommodate partial changes. Now that furniture arrangements, interior partitions, wiring layouts, etc., can all be constantly compared with the latest construction floor plan and nothing other than work affected by changes need be redrawn, it's very easy to keep all work thoroughly up to date in relationship to other work.

You'll find a similar advantage with complex site work. It used to be that contractors trenching for drainage might be out of sync with others trenching for gas or electrical work. When this happens, you have situations where one subcontractor accidently damages existing or newly placed work of another contractor. Or a trench is dug, work completed, the trench filled and paved over only to have to be redug for other work that happens to follow the same line.

The trick to letting everyone on the job know what's happening is to create all site work drawings as overlays and print each separate utility line or contract or discipline as a solid-line image on a screened image that shows all the other work. In the past the various consultants that might lay out site work couldn't be expected to show the work of other consultants on their drawings. So it was up to the prime designer to check for conflicts, and it was up to the prime contractor to ensure coordination, with inevitable mixed results. By using overlays and screening techniques you can combine any number of trades on one drawing and screen or not screen any of them. These composite reference prints show everybody exactly how their work relates to the work of others and how to avoid duplication of trenching, interferences, bad crossovers, and inadvertent damage to other work.

MOCK-UPS

It's clear how coordination and checking are expedited through the overlay drafting and printing processes. Another great asset derives from the fact that a systems job has to be thoroughly planned and mocked up in miniature before you start drawing production. That requires making a list of all drawings, as well as lists of base and overlay sheets, and making sketch cartoons of final drawing sheets to show approximate sizes and locations of all major components of each final drawing. When using overlay drafting, the project manager must also sketch miniature base and overlay sheets. If this first stage of coordination is completed, it helps ensure coordination throughout the job. If it isn't done, coordination management becomes increasingly difficult through the day-to-day distractions and brush-fire fighting of typical traditional crisis administration.

STOP POINTS

The key concept for successful coordination of base and overlay data between all job participants is the idea of the stop point. All work that is to be used by any other discipline has to be brought up to a point of completion as suitable for reference use by others, stopped, and printed in base sheet form. Thus, as described in Chapter 6, architectural plans are to include what's needed by electrical, plumbing, and HVAC consultants and then are stopped; all further work—including elaborations on architectural data—is done on overlays. This point has to be clarified: any revisions in architectural plan are *not* done on overlays unless they're tentative or created as options or alternatives. Revisions to the architectural plans are done on the *original* base sheet(s), not the copies that people are using with their overlays. The architectural work that *is* done on overlays consists of the layers of data that go beyond the stop point, data *not* needed by other consultants. Such data include strictly architectural notes, door symbols, finish schedule symbols, detail bubbles, etc.

When original base sheet information is revised and updated, memos, slip sheets, and, when necessary, new base sheet copies are submitted to all participants.

The reference sheets in Figures 8-1 through 8-5 from the northern California chapter AIA publication *Recommended Standards on Production Procedure* show the kinds of data normally included at the stop point for various consultants.

COORDINATION OF CONSULTANT DOCUMENTS

CONSULTANT	ARCHITECTURAL DRAWINGS	CONSULTANTS' DRAWINGS
STRUCTURAL	1. Grid lines, numbers and grid dimensions. Floor elevations shall be top of structural floor.	1. Grid lines, numbers and grid dimensions. Floor elevations shall be top of structural floor.
	2. Outline of foundation in sections and details only.	2. Foundation plan with all dimensions and elevations of bottom of footings.
	3. Only outline structural elements where in conjunction with architectural work. Avoid necessary repetition of structural items.	3. Locate and size all structural elements.
	4. Locate and dimension all architectural openings in structure, slab depressions, raised floor and curbs.	4. Show all openings and critical penetrations in structure graphically, locate if critical and coordinate with other consultants. Indicate depth of floor depressions and integral raised floors and curbs.
	5. Refer to "See Structural Drawings" only when architectural item is defined by structural drawing. Do not refer to specific structural detail.	5. Refer to Architectural Drawings only in General Notes.
	6. Locate and dimension all work outside of building such as sidewalks, patios, stairs on grade, trenches, etc.	6. Typically detail structural elements of all work outside of building. Show graphically outline only where in conjunction with structural elements.
	7. Detail Miscellaneous Metal, including metal stairs. (Obtain structural sizes and connections from consultant).	7. Show connections of all miscellaneous metal to structural elements. Submit to Architect and be responsible for all sizes and connections.
	8. Show below grade membranes (below slab and on walls) in sections and exterior elevations.	8. Show below-grade membranes graphically only where in conjunction with structure.
	9. Fire proofing to structural steel shall be shown on all details where in conjunction with architectural work. (Thickness and rating is established in Specification)	9. For the purpose of establishing the required applied fire proofing to structural steel and decking, identify girders, beams, columns, etc., as to being primary, secondary, etc., members.
	10. Show elevator shaft and elevator machine beam support. Locate machine beams supplied by Elevator Manufacturer.	10. Detail size and locate all elevator shaft and machine beam supports.

Fig. 8-1. (Courtesy of the northern California chapter, American Institute of Architects.)

COORDINATION OF CONSULTANT DOCUMENTS

CONSULTANT	ARCHITECTURAL DRAWINGS	CONSULTANTS' DRAWINGS
MECHANICAL	1. Provide background of site and building plans. If working transparency is supplied, room names and numbers shall be indicated on separate print.	1. Background to have grid lines and numbers, all structural elements, partitions, door swings and pertinent built-in work, room names and room numbers. Background shall be maintained to reflect the final architectural and structural work.
	2. Do not show any mechanical elements unless contiguous with an architectural item or part of the architectural appearance and design concept.	2. Identify all mechanical items, coordinate with other consultants. Show sizes of penetrations through structure, locate graphically only.
	3. In finished spaces show outline of equipment only. Show equipment bases and curbs where not part of the mechanical work.	3. Locate all mechanical items and detail equipment bases, curbs, catwalks, concrete pits, roof penetrations, flashings to equipment. Coordinate all reinforced concrete requirements with Structural Consultant. Where integral with structure of building submit requirements to Structure Consultant.
	4. Locate by dimension non-typical exposed mechanical items for critical architectural appearance.	4. Locate, dimension and size all typical or standard mechanical items. Show all required access panels. Verify equipment connections to structure and floor loading requirements with Structural Consultant. Coordinate mechanical work with other consultants for clearances.
	5. Reflected ceiling plan shall show outline of all exposed mechanical items. Reflected ceiling plan, floor plans, elevations and selections to clearly indicate changes in floor and ceiling levels.	5. Show location of floor and ceiling level changes which affect mechanical work (for clearances) and refer to architectural drawings for specific conditions.
	6. Indicate increased wall or partition thickness for mechanical work. Dimension clear openings of shafts.	6. Submit to Architect space requirements for all enclosed mechanical work.

Fig. 8-2. (Courtesy of the northern California chapter, American Institute of Architects.)

COORDINATION OF CONSULTANT DOCUMENTS

CONSULTANT	ARCHITECTURAL DRAWINGS	CONSULTANTS' DRAWINGS
PLUMBING/SPRINKLER	1. Provide background of site and building plans. If working transparency is supplied, room names and numbers shall be indicated on separate print.	1. Backgrounds to have grid lines and numbers, all structural elements, partitions, door swings and pertinent built-in work, room names and room numbers. Background shall be maintained to reflect the final architectural and structural work.
	2. Do not show any plumbing elements unless contiguous with an architectural item, or part of the architectural appearance and design concept.	2. Identify all plumbing items, coordinate with other consultants. Show sizes of penetrations through structure, locate graphically only.
	3. Show outline only of all fixtures including floor drains, locate nontypical by dimension.	3. Indicate all plumbing items, give dimensions of all typical locations and standard installations. Show sizes of penetrations through structure, locate graphically only. Coordinate with other consultants. Show all required access panels. Verify plumbing work connections to structure and floor loading requirements with Structural Consultant.
	4. Indicate sprinkler head locations on reflected ceiling plan for critical architectural appearance.	4. Show location of sprinkler valves and all sprinkler mains. Coordinate with other consultants for clearances. Where required for architectural appearance refer specifically to architectural drawings for sprinkler head locations.
	5. Show lcoations only of fire hose cabinets, fire extinguisher, etc. (to be specified under Plumbing).	5. Locate, size and identify fire hose cabinets, fire extinguisher cabinets, fire extinguisher etc. (also specify).

Fig. 8-3. (Courtesy of the northern California chapter, American Institute of Architects.)

COORDINATION OF CONSULTANT DOCUMENTS

CONSULTANT	ARCHITECTURAL DRAWINGS	CONSULTANTS' DRAWINGS
ELECTRICAL	1. Provide background of site and building plans. If working transparency is supplied, room names and numbers shall be indicated on separate print.	1. Background to have grid lines and numbers, all structural elements, partitions, door swings and pertinent built-in work, room names and room numbers. Background shall be maintained to reflect the final architectural and structural work.
	2. Do not show any electrical elements unless contiguous with an architectural item, or part of the architectural appearance and design concept.	2. Identify all electrical items, coordinate with other consultants. Show sizes of penetrations through structure, locate graphically only.
	3. Show electrical panels, surface or recess mounted, on plans. Check partition depth and possible necessary furring.	3. Define and locate all electrical panels. If necessary indicate all critical clearances around panels and equipment. Verify location of oversize conduits or ducts with Structural Consultant if imbedded in structure and with Architect if critical for architectural appearance and coordination. Detail all reinforced concrete work required for electrical items and coordinate with Structural Consultant, such as trenches, duct banks, concrete pits, equipment bases, etc.
	4. Reflected ceiling plan shall show all ceiling fixtures and other electrical items. Locate or indicate by module for critical architectural appearance.	4. Locate graphically all electrical items, coordinate with other Consultants.
	5. Locate by dimension non-typical electrical wall outlets and fixtures only if critical for architectural appearance. If locations of floor outlets are critical, issue a "furniture plan" with dimensions for floor outlets.	5. Locate and dimension outlets and fixtures by typical reference only. Refer to architectural drawings for non-typical locations of all electrical items.

Fig. 8-4. (Courtesy of the northern California chapter, American Institute of Architects.)

COORDINATION OF CONSULTANT DOCUMENTS

CONSULTANT	ARCHITECTURAL DRAWINGS	CONSULTANTS' DRAWINGS
CIVIL	1. Architectural Site Plan to show outline of all new and existing structures, roads, paths, appurtenances and any other architect-designed items. Indicate all finish elevations of the before-mentioned items including entrance platforms, top of area ways, etc.	I. Show outline of all architect-designed items. Indicate bench marks, property lines, extent of all new site work.
	2. Omit from architectural site plan all work covered by civil engineers' drawing except as stated above.	2. Indicate extent of cut and fill, rough grading, all hard surfaces, curbs, walls, retaining walls, catch basins, slope of drainage and all required finish grades and elevations.
FOOD SERVICE	1. Show outline only of large built-in equipment (walk-in refrigerators, etc.). Indicate slopes of floor. Determine all room finishes except for pre-finished equipment.	1. Background plans to have grid lines and grid numbers, all structural elements, partitions, door swings, room names and numbers. Indicate and detail all items supplied and/or installed under Food Service Equipment.
	2. Reflected ceiling plan shall show outline of all food service equipment extending to the ceiling. Also show outline of hood and other ceiling penetrations.	2. Show all ceiling penetrations. Coordinate with other Consultants al Food Service equipment. Provide a complete curb and setting plan for all equipment. Provide complete rough-in data for all mechanical, plumbing, sprinkler and electrical connection indicating sizes and loads required.
LANDSCAPING	1. Clearly define outline of landscaped areas. Show outline of all new and existing structures, roads, paths, appurtenances and any other architect-designed items. Indicate all finish elevations of the before-mentioned items including entrance platforms, top of area ways, etc.	1. Show outline only of all architect-designed items. Show clearly extent of landscaped area, indicate completely and detail all items furnished under landscaping. Show sprinkler, irrigation and drainage including tie-in to plumbing. Locate irrigation controller.
	2. In elevations and sections show existing grade as well as finish grade.	2. Indicate clearly extent of cut and fill.
	3. Hard surface drainage.	3. Coordinate with other Consultants and locate by dimension all items which are critical to landscaping (light standards, water supply for fountains, retaining walls, etc.)

Fig. 8-5. (Courtesy of the northern California chapter, American Institute of Architects.)

9

PLANNING OVERLAY DRAFTING AND GRAPHICS

THE SIMPLE FAIL-SAFE OVERLAY PLANNING SYSTEM

Planning is everything with overlay, but the steps of planning an overlay project have seemed arcane and so mysterious that many architects and engineers won't do it or won't do it adequately. Instead they pass the job on to others, cross their fingers, knock on wood, take a trip, and hope for the best. Here are the steps. As you'll see, it's not that big a deal.

FIRST PHASE— SCHEMATICS TO DESIGN DEVELOPMENT

1. List the architectural drawings that go into the project.

2. List consultants' drawings required, if any.

3. If the design development or presentation drawing package is very complex, make a miniature mock-up with sketches or cartoons of the data that go on each sheet.

4. After doing the index of final drawings, make a list of drawing separations or layers required to make each final drawing. The rule is that you don't want to "use up" any original drawing that can be used in some different context. A floor plan, for example,

might be shown in phased construction. The core floor plan will then be the base sheet with the phases or variations in plan being done on overlays.

5. Under each listed design development or presentation drawing list the base sheet and the overlay(s) that will comprise that final drawing. For example, a furniture plan will be made up of the plain line work of the floor plan and an overlay or overlays showing the different furniture plans.

One goal in all this is to keep all extraneous data off of the original floor plans, elevations, site plans, and sections. Do all notes, titles, rendering, etc., solely on overlays. Then, to whatever degree those original drawings are reusable in working drawings, they can indeed be reused as base sheets instead of being drawn all over again.

SECOND PHASE— WORKING DRAWINGS

1. Continue the listing process. Make a legend or index of all anticipated final architectural drawings. Allow ample space under each listed drawing to make a sublist of layers that might make up any final drawing. And allow space for later additions as the drawing list is revised over time.

2. Make a list of anticipated final consultants' drawings in the same fashion as above.

3. If the job is complex, circulate the drawing

83

lists to others who are concerned with the job and get their input on what drawings might be added or omitted.

4. When the list is reasonably complete, make a miniature mock-up of sketches or cartoons of the final sheets. You might be able to use photocopy reductions of design development drawings as starters for the mock-up set.

5. The mock-up will show you constants and variables applicable to base and overlay separations. If you see the floor plan repeated on sheet after sheet of consultants' drawings, you have the clearest and most obvious constant and variable relationship.

6. Examine the mock-up for other large repetitions suited to base-overlay separation and make a sublist of likely base and overlay layers as part of your list of final drawings.

7. Just as each final drawing has an identifying sheet number, so all base and overlay drawings must have individual code numbers. Note these numbers with the sublist of drawing layers.

8. Make job policy decisions on sheet separations. I recommend keeping the layers minimal, especially in early efforts at overlay. Keep in mind that overlays apply only to a portion of the job, mainly floor plans, and not to all drawings. There's no great gain in using a title block and border layer as a base sheet for every single drawing in lieu of just having them preprinted on final sheets. Note that most aspects of repetition are best handled by paste-up composite drafting. It's large-sheet elements that are most suited to the overlay system. Decide if you'll use a separate building grid base sheet, or separate room name intermediate overlays, etc. *For most final drawings you'll really only need one base and one overlay, one general constant (such as the architectural floor plan) and one general variable (such as the HVAC).* Typically, have the title block on the base sheet and the final drawing name and number on the overlay.

9. Identify the stop points of drawing—the amount of information needed and not needed when making copies for base sheets. There'll be a stop point in design development when plain raw-building line work is complete; all variations and rendering elements are then finished off on overlays. There'll be a stop point further on when the architectural drawings have data needed by the consultants: door swings for the electrical engineer, for example, so light switching can be planned; plumbing fixtures and related data needed by the mechanical engineer. Drawings are stopped, base sheets made and distributed, then everyone—architectural and engineering drafters—continues the work on overlays. The architectural staff add in notes, dimensions, door symbols, detail keys, etc., on their overlays. Engineering drafters proceed with their independent work on their overlays as indicated in the miniature working-drawing mock-ups. (See Chapter 8 on coordination among consultants for more elaboration on this process.)

10. Make a supplementary mock-up. The first working-drawing mock-up is a miniature of the final set of drawings. The supplementary mock-up is a miniature of bases and overlays that come together to make the final set. This can be a partial mock-up just to clearly show yourself, the consultants, and drafting staff the actual physical relationships of these sheets both in drafting and later in printing. This is the place to resolve problems, make policy decisions on how far to go with layering, and, most important, clear up the misunderstandings that so often destroy the efficiency and economy of the overlay system.

11. On the smallest jobs, the single index and subindex of bases and overlays will suffice for planning. On medium to larger jobs, you'll need the planning indexes and mock-up and one or two more planning and management tools. One of these is the print control matrix. This is a chart, as shown in Figures 9-7 and 9-8, that shows all final drawings listed on one side and all bases and overlays listed on the other. Marks in the field of the matrix show which bases and overlays must be combined to make up a final total print.

12. The final overlay management tool is the print control sticker, also called a "match box," "ladder," "sticker matrix," among other names. This is necessary for more complex jobs as a final check in the compositing of drawings when setting up to print and later when checking the prints.

The information for the stickers is taken from the print control matrix described in step 11. This, like the other steps, is to be done by the project manager. It's to be done, along with the print control matrix, prior to the first interim check printing.

As will be obvious, you couldn't very well sit down and do the print control stickers or print control matrix all by themselves. The information that goes on these final overlay control devices is derived step by step from the drawing list, the drawing mock-up, and the sublist. If you follow the steps as outlined, it will all be a logical and fairly self-evident process. If you don't follow the steps and don't have a truly comprehensive overlay knowledge and experience, the job will become hopelessly confused and muddled. The illustrations on the next few pages provide a visual review of the steps just listed.

GRESHAM AND SMITH
INDEX OF DRAWINGS

SHEET NUMBER	SHEET TITLE
	Cover Sheet
	Legend Sheet (inside cover if offset)
C·1.1	Geometric Plan
C1.2	Grading & Drainage Plan
C3.1	Civil Details
L1.1	Landscape Plan
L1.2	Courtyard Plan
L3.1	Landscape Details
A2.1	First Floor Plan
A2.2	Second Floor Plan
A2.3	Roof Plan
A3.1	Elevations
A3.2	Sections
A3.3	Details
A3.4	Details
A4.1	Large Scale Plans
A5.1	Ceiling Plan
A5.2	Ceiling Plan
A6.1	Elevator / Stair Details
A7.1	Misc. Details
G1.1	Graphic Site Plan
G2.1	First Floor Graphics
G2.2	Second Floor Graphics
D2.1	First Floor Furniture and Accent Walls
D2.2	Second Floor Furniture

Fig. 9-1. Step 1—the list of drawings. List the architectural drawings and all the consultants' drawings in a preliminary "table of contents."

(Courtesy of Ed Powers, Gresham, Smith and Partners.)

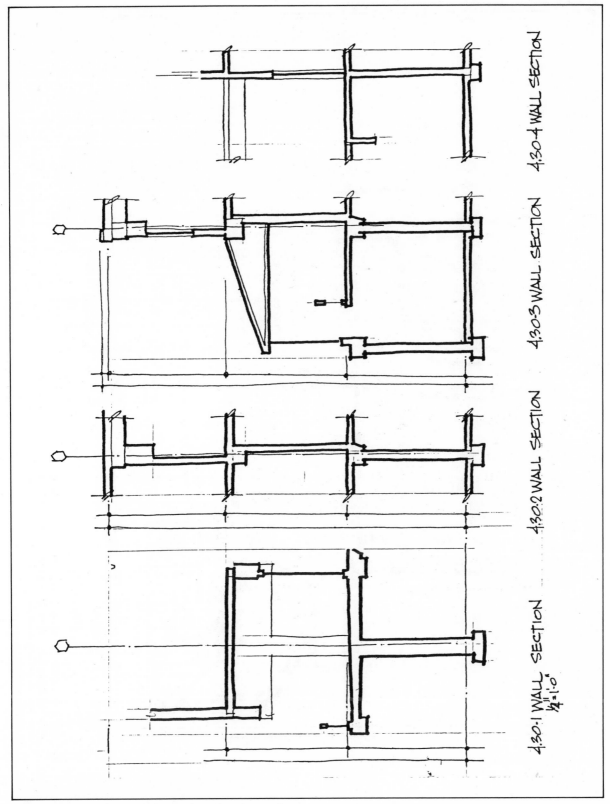

Fig. 9-2. Step 2—the miniature working-drawing mock-up sheets. Contents of each sheet are sketched or cartooned to about this degree of completeness, mainly enough to show what goes where on each sheet and about how much space it will take.
(Illustration turned on page.)

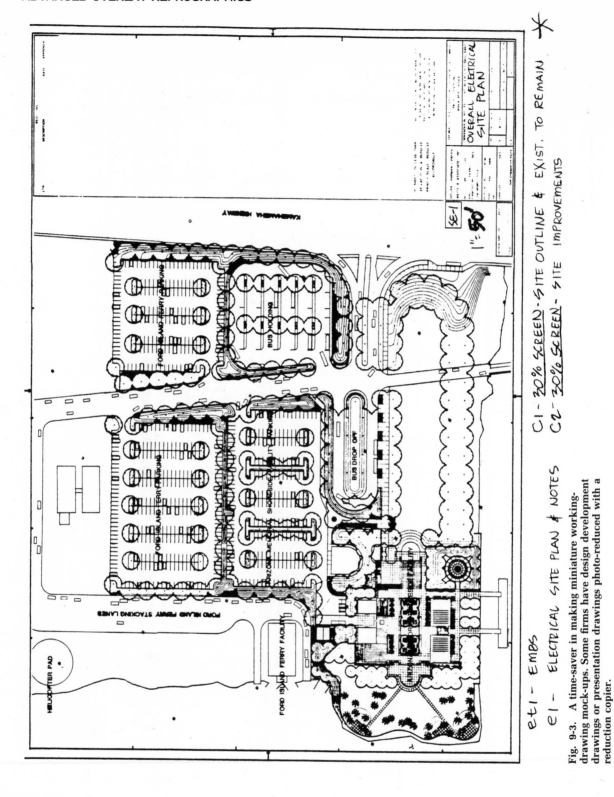

Fig. 9-3. A time-saver in making miniature working-drawing mock-ups. Some firms have design development drawings or presentation drawings photo-reduced with a reduction copier.

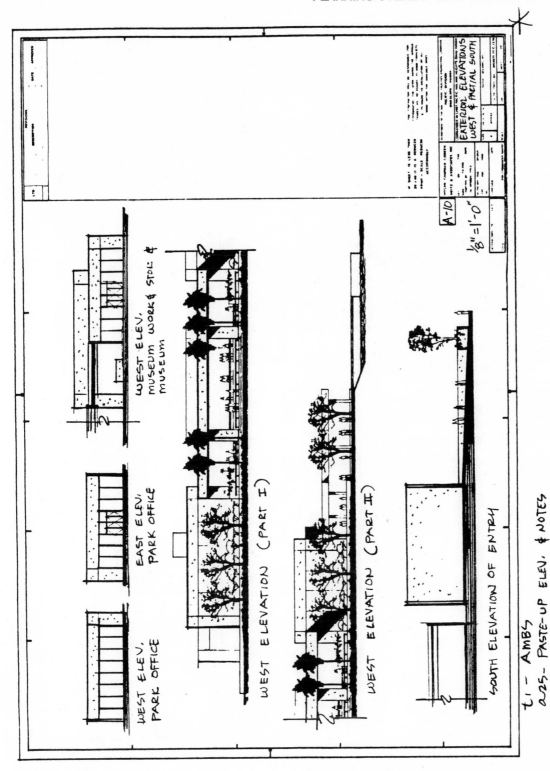

Fig. 9-4. Photo reductions are made of elevations and sections as well as plans for the mock-up.

sht. no.	sheet description	overlay/base no.	screen req'd (%)	assign	remarks * = 30% SUBMITTAL
C-16 *	SITE UTILITIES PLAN - SECTOR II 1"=20'	ct1			CIVIL MASTER BORDER SHEET
		C6	30%		SITE OUTLINE & EXISTING TO REMAIN FOR SECTOR II
		C12	30%		SITE IMPROVEMENTS (ALL NEW WORK) FOR SECTOR II
		C42			WATER PLAN & NOTES FOR SECTOR II
		C48			SEWER PLAN & NOTES FOR SECTOR II
		C54			STORM DRAIN PLAN & NOTES FOR SECTOR II
C-17 *	SITE UTILITIES PLAN - SECTOR III 1"=20'	ct1			CIVIL MASTER BORDER SHEET
		C7	30%		SITE OUTLINE & EXISTING TO REMAIN FOR SECTOR III
		C13	30%		SITE IMPROVEMENTS (ALL NEW WORK) FOR SECTOR III
		C43			WATER PLAN & NOTES FOR SECTOR III
		C49			SEWER PLAN & NOTES FOR SECTOR III
		C55			STORM DRAIN PLAN & NOTES FOR SECTOR III
C-18 *	SITE UTILITIES PLAN - SECTOR IV 1"=20'	ct1			CIVIL MASTER BORDER SHEET
		C8	30%		SITE OUTLINE & EXISTING TO REMAIN FOR SECTOR IV
		C14	30%		SITE IMPROVEMENTS (ALL NEW WORK) FOR SECTOR IV
		C44			WATER PLAN & NOTES FOR SECTOR IV
		C50			SEWER PLAN & NOTES FOR SECTOR IV
		C56			STORM DRAIN PLAN & NOTES FOR SECTOR IV

Fig. 9-5. Step 3—by analyzing the miniature working-drawing mock-up for constants and variables, it's possible now to sort out layers of base and overlay sheets. This is a partial drawing list with a supplementary list of layers that make up each final printed sheet in the set.

OVERLAY ORGANIZATION SHEET

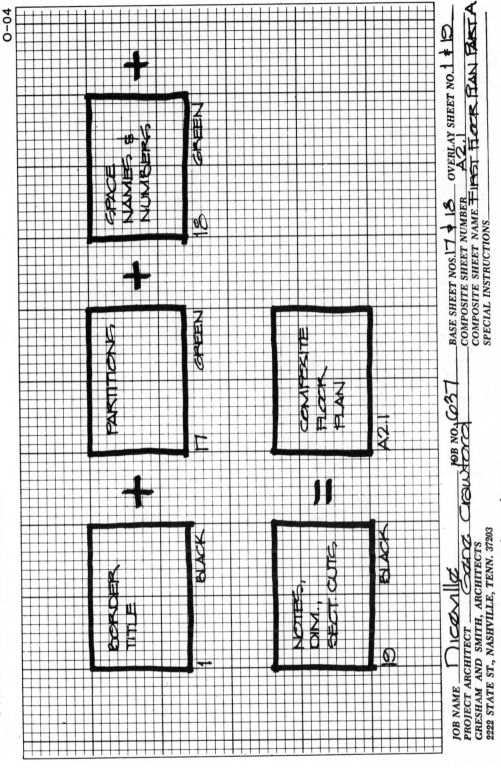

JOB NAME Dicenlle ___ *JOB NO.* 637 ___ *BASE SHEET NOS.* 17 & 18 ___ *OVERLAY SHEET NO.* 1 & 19

PROJECT ARCHITECT Gary Crawford ___ *COMPOSITE SHEET NUMBER* A2.1

GRESHAM AND SMITH, ARCHITECTS ___ *COMPOSITE SHEET NAME* First Floor Plan Parta

2222 STATE ST., NASHVILLE, TENN. 37203 ___ *SPECIAL INSTRUCTIONS*

Fig. 9-6.   Here's another way of showing base and overlay sheet separations.

(Courtesy of Ed Powers, Gresham, Smith and Partners.)

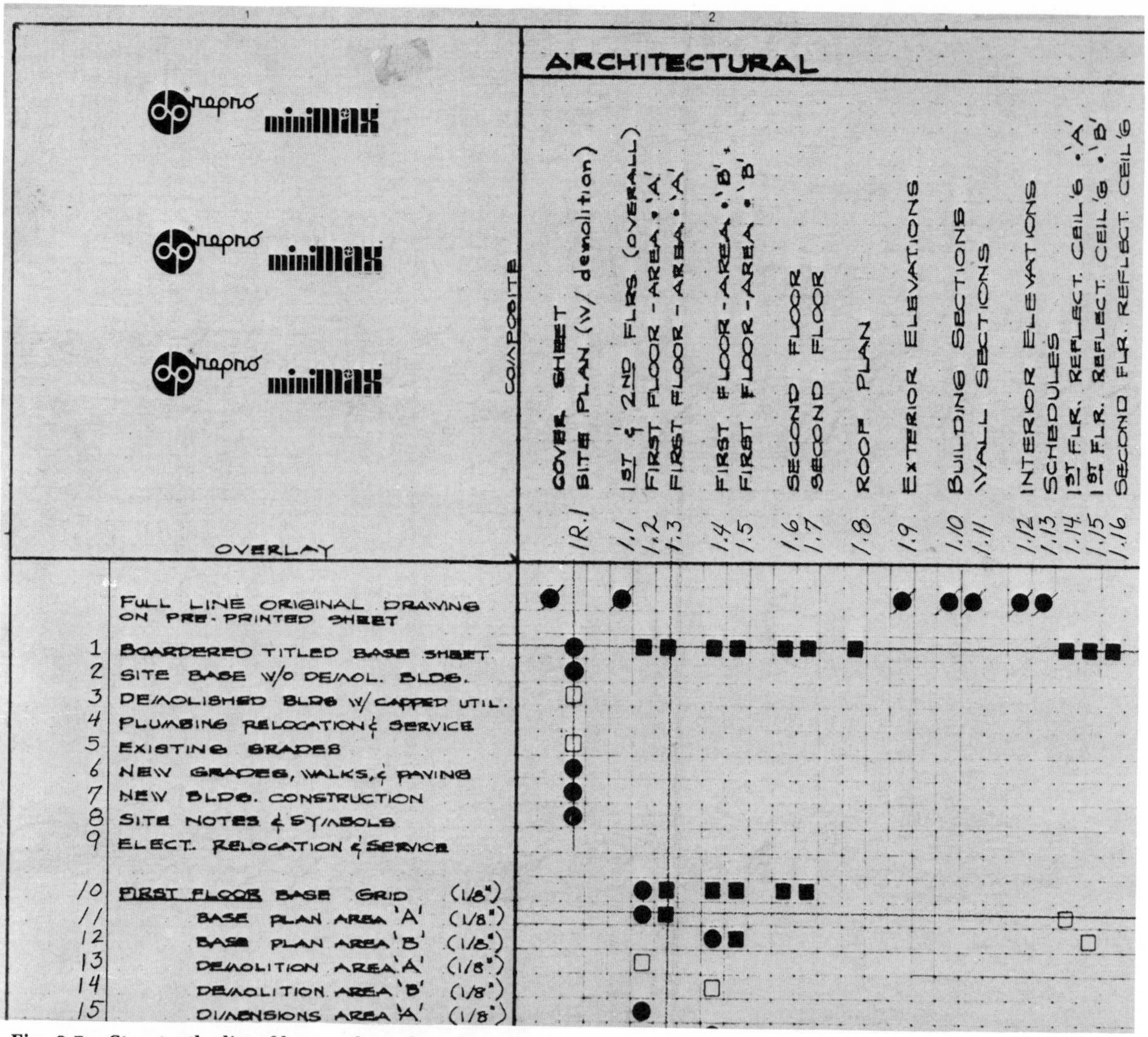

Fig. 9-7. Step 4—the list of base and overlays shows how to make up a print control matrix. That's a chart listing all bases and overlays on one side, all final prints on the other, with marks in the field of the matrix to show which bases and overlays combine to make which final prints.

(Courtesy of miniMax.)

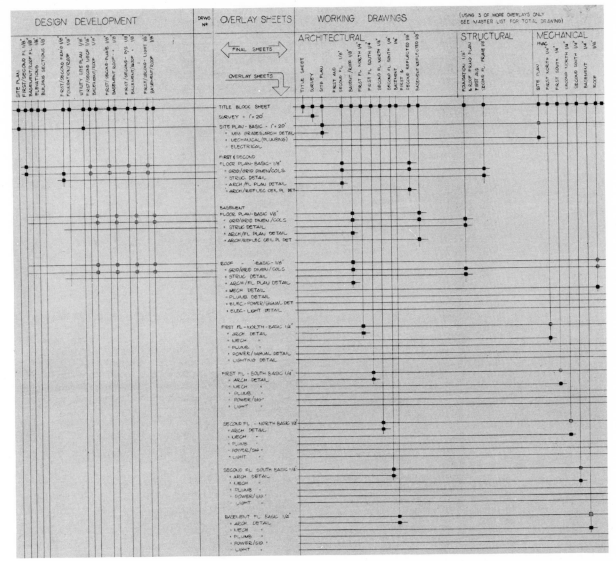

Fig. 9-8. Another version of the print control matrix. All such charts follow the same principle but vary in details of graphic design.

(Courtesy of miniMax.)

Fig. 9-9.   Step 5—the final tool in controlling the correct sequencing when printing base and overlay sheets is the print control sticker. It starts with a preprinted stickyback two- or three-column grid placed on the job captain's layout board. That becomes a template to guide precise identical placement of all subsequent stickers on all the base sheets.

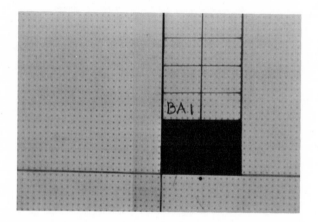

Fig. 9-10.   This is a close-up of the sticker as it looks when printed on a sepia line slick copy of a base sheet. This base sheet is numbered BA1 and is so noted in the left-hand lower box. The base sheet normally has the title block with job and office name on it, so separate job identification isn't necessary in this case.

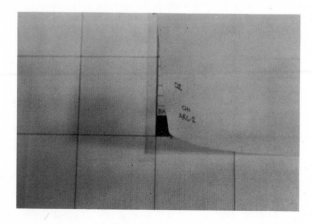

Fig. 9-11.   The next layer overlay is placed in registration with the base sheet. Its sheet number is marked so as to fall within the second-level box on the print control sticker image on the base sheet.

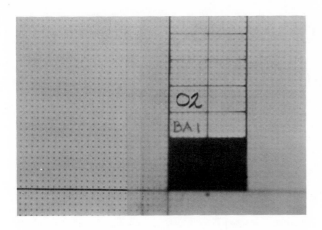

Fig. 9-12.   This is a close-up of the overlay sheet atop the base sheet. Its number is O2. Since it's an overlay sheet, it will have some coding to identify the office or project name or both. Without that coding, the sheet can easily be misplaced and may be difficult to identify later. The coding is written in where it will be blotted out by the mask-out rectangle. You can see coding in both the preceding and the next illustration.

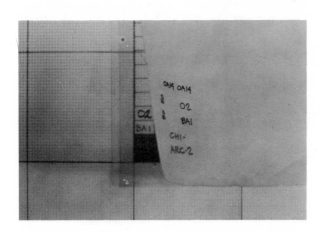

Fig. 9-13.   The third and final layer is put in place. Every sandwich of base and overlays always has a layer which is unique or has the most variable data. The other layer or layers have data that are common to other combinations. The most unique overlay is the one used only once in the job. It has the final drawing title and the final sheet number which will fall in place within the title block that's on the base sheet. It also has the final list of all layers that are combined to make up the final sheet.

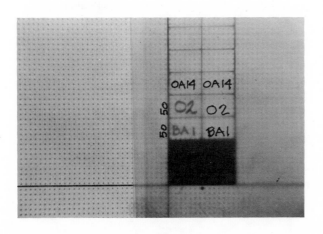

Fig. 9-14.   This is a close-up of the final overlay drawing list that falls within the right-hand column of the print control sticker. When the planning and mocking up of this project was done, each sheet was identified according to the base and overlay combinations required to make that sheet. The list is on the overlay drawing index and the print control wall matrix. That same list is now shown on the final overlay.

The final overlay code number is OA14. That number is written in both columns across the top of the lattice. The intermediate overlay and the base sheet are numbered below it as shown. The percentage of screening of base and intermediate overlay is noted on the left-hand side.

This is the final visual check. If a sandwich of sheets is made for printing and there's a number missing in the left-hand column, it means there's a sheet missing. If there's a number up in space somewhere, it means there's a sheet that doesn't belong. The numbers provide the final guide for the printer, and the final check after printing.

FLOW CHART FOR DRAWINGS
PRODUCED BY TEAM DRAFTING

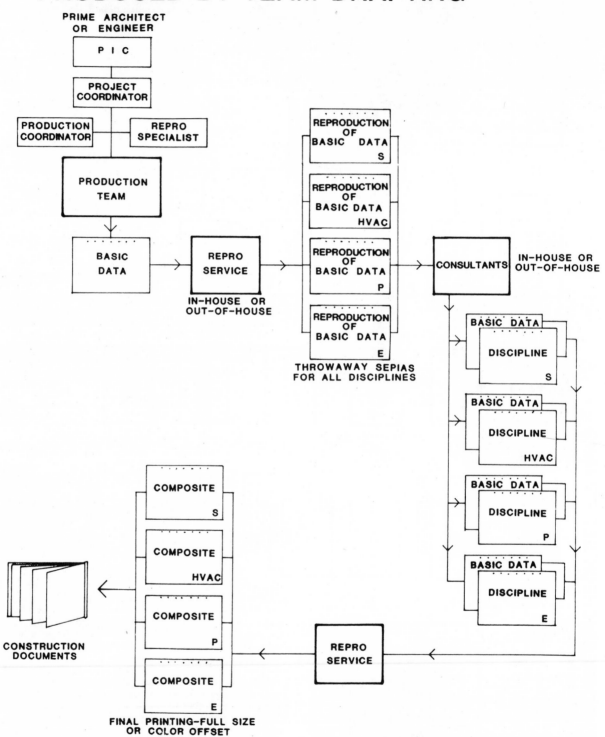

Fig. 9-15.   (Courtesy of Ed Powers, Gresham, Smith and Partners.)

96

PROGRESS/CHECK SET PRINTING

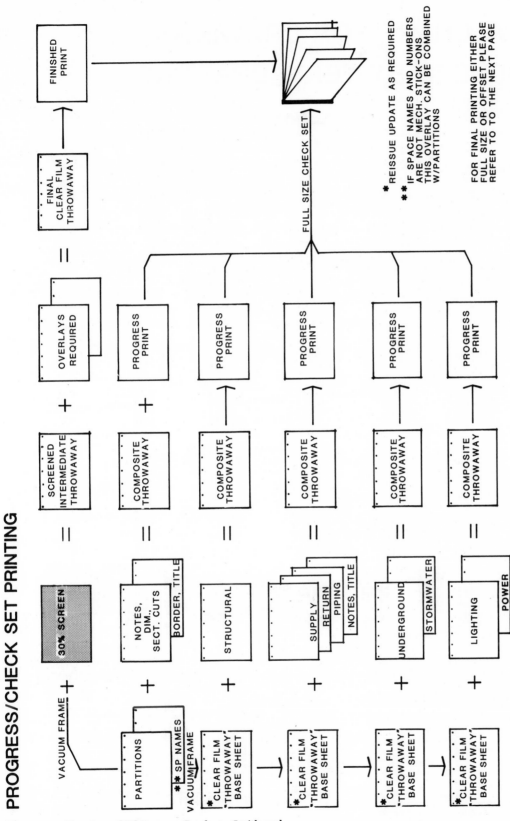

Fig. 9-16. (Courtesy of Ed Powers, Gresham, Smith and Partners.) *(Illustration turned on page.)*

CONSTRUCTION DOCUMENT PRINTING

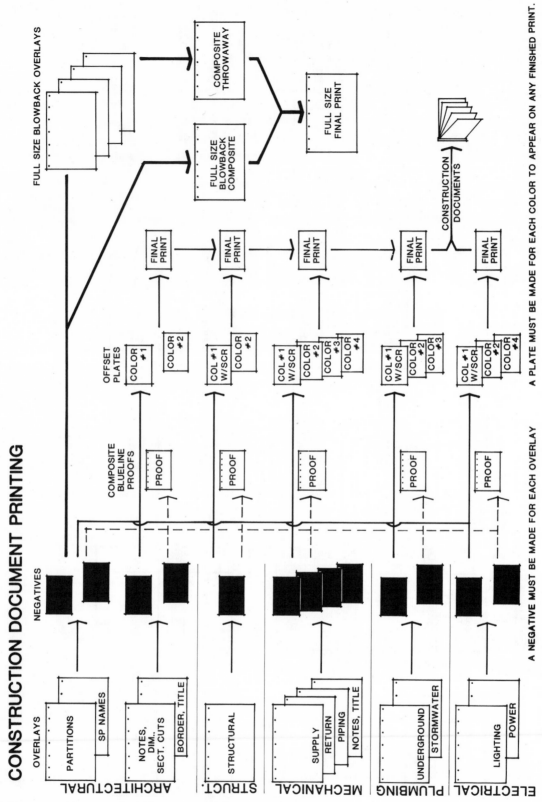

Fig. 9-17. (Courtesy of Ed Powers, Gresham, Smith and Partners.) *(Illustration turned on page.)*

PART 3
COMPUTERIZING ARCHITECTURAL AND ENGINEERING PRACTICE

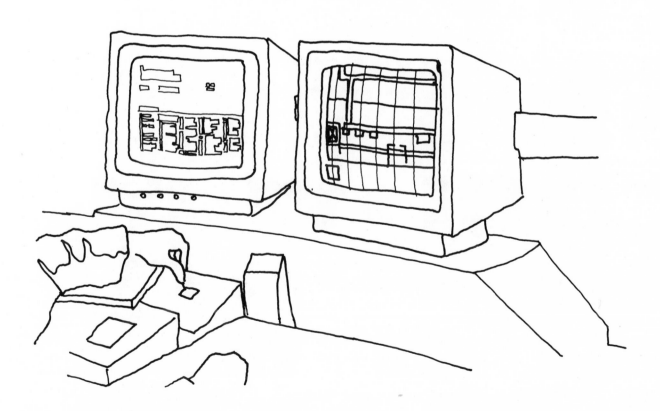

Illustration by Lanée La 'shagway.

10

COMPUTERIZATION AND COMPUTER GRAPHICS

INTRODUCTION

I'm writing this book on a computer terminal. I'm using a reasonably high-powered desk top microcomputer and a simple but versatile word processing program. I also have access to a program which will check my spelling, hyphenation, and even some aspects of my writing style.

When doing later books, I'll send my manuscript from my studio computer out here in the woods of Orinda, California, via telephone to the publisher's computer in another city. Their editors will send revised copy back to me the same way. Ultimately, I'll be able to send final copy back in such a form that it can be fed directly into the computer that controls their typesetting machine.

In other words, the old process that required two to three different steps of typing, retyping, and then typesetting, plus the mailing of manuscripts back and forth, will be transformed entirely.

So will the writing process. I have just written a paragraph which I've decided belongs at the end of this introductory section. I don't have to eliminate and rewrite the paragraph. Instead, I punch in the line numbers and give the computer a "MOVE" command, and the paragraph is now moved to the end of the text. And when I rewrite the paragraph to change a word here and there, or make corrections, or remove a sentence, it's all no more than punching the keys on the keyboard. It's the electronic version of paste-up. It's the electronic way of exploiting reusability.

In the traditional mode of manual typing, every time you made a major change, you had to change everything else that went with it. To move a paragraph might mean retyping a couple of pages. If there were enough changes to be made on a page, it meant retyping the page. That's been the whole problem with nonsystems drafting and graphics. Potentially reusable drawing or notation was always mixed in with the nonreusable in such a way that reusability was too much trouble. It was just as fast and easy to redraw everything in its new context.

That's the beauty and the power of computerization: ease of creation of original, potentially reusable data, convenient storage of reusable data, ease of manipulation of that data, and convenient retrieval. That's what systems is all about, and computers are the quintessence of systems.

COMPUTERS IN THE ARCHITECTURAL AND ENGINEERING PROFESSIONS

The impending architectural and engineering changeover to all-out computerization is only part of a total nationwide cultural, technical, and communications revolution that's affecting everyone in every kind of job.

Most large-scale businesses adopted computers years ago. Now smaller companies, professional offices, and even one-person enterprises are switching over. Many thousands of individuals own personal computers now and bring them to work much as people used to bring in their pocket calculators. Everyone in every field is facing the necessity of changing, and although it requires the discomfort of altering some habits and the tedium of learning new skills, the change is inevitable.

Studies by Harry Mileaf of Sweet's on architectural and engineering computerization indicate that 35 percent of architectural firms and 84 percent of engineering firms had computers or advanced calculators in use as of 1982. By 1986 that will change to 79 percent of architectural firms using computers and 95 percent of small- and medium-sized engineering firms. (All large engineering firms are already well involved in computerization.)

Whereas well under 1 percent of architectural and engineering firms now use computer graphics and automated drafting (computer-aided design and drafting), that percentage is likely to change to 20 percent over the next four years, and double and redouble again by 1989.

These percentages of change will be modified up or down and the time factor will change depending on recessions, war, and other political interventions. But anyone who fails to plan to computerize as soon as it's demonstrably cost-effective is heading for serious problems.

It takes time to implement a computerization program of any size—at least a year for the major steps of the process. So you will see some offices start the process long after others are old hands at it. Then they will require a year to get things working—and their competitors will have that same year to accelerate their advances. In other words some firms will start too late and eventually give up and just fold.

People, on an individual basis, will have the same problem. First, those versed in computer practice will command larger salaries. Those not versed will find their skills have less value in the job market. At some point the amount of catching up required will be more than many people want to endure, and they will drop out.

All that's needed to make a successful transition is information: how the machines work and how you can most readily put them to work for you. That's one of the functions of this book. With Guidelines manuals, my new books and software, and a flood of other literature and training programs to come, there'll be no shortage of the information you need to take maximum advantage of the changes to come.

WILL THE COMPUTER TAKE YOUR JOB AWAY?

The first thing to know is that computers won't take away your job or make your job or anyone else's job less important. Just the opposite will happen.

How can that be? A computer supposedly does more work faster and cheaper than people can, doesn't ask for raises, and works nonstop without complaining. How can it possibly be that computers won't replace or displace some employees?

What computers do is accelerate work flow; that is, they help move more work through the firm faster. The machines don't do the work faster in themselves; they help people do the work faster, and they help people do more varieties of work. That increases—rather than decreases—the employees' job roles.

Computers boost the marketing process by helping the firm reach more prospective clients in shorter time. They help in preparing job proposals and, as a feature of the office's services, are attractive to clients. Computers improve response time in dealing with clients, and they add new varieties of design and management services. That too helps keep new work coming in to keep the firm alive.

Computers create new work. Since they can do some tasks that wouldn't be economical to do otherwise, such as certain complex solar or energy conservation analyses, those tasks are added to office functions. That means work is added and, side by side with the computer, someone has to be around to do it.

So rather than taking work away, the computer, like any tool, changes work, augments it, and increases its value by allowing you to do more of it in less time. Here are examples of what happens to various job roles:

- The typist does more routine typing faster through word processing on a computer and even spells and punctuates better with the aid of special editing programs. So that part of the work is accelerated and reduced in time. The typist can then take on other jobs of storing reusable wordings and information in the computer memory. Later he or she can assemble and rearrange stored data and create new proposals to prospective clients, press-release background data, brochure data, forms for government agencies, mailing lists to publicity sources and client prospects, etc.

- The spec writer writes specifications faster with prewritten, computer-stored outline-specification sections. The computer reduces the labor, leaving more time to review new products and technical data. Now the spec writer has time and equipment to ensure more exacting coordination between specifications and working drawings. The specification writer does less of some kinds of work but now can do more research and thought work.

- The administrator gets more information to improve accuracy in project scheduling and budgeting and gets it faster. That opens up time to monitor more aspects of the office's functioning and allows for more thorough and better-researched long-range office planning.

- The designer receives computer-generated schematic floor plans that are exactly accurate to the client's stated spatial programming needs. That cuts the false starts and endless revisions common to the traditional schematic and design development process. With time saved in creating the background information and schematics, the designer can spend more time reviewing variations and refinements so the design that goes to production represents a more complete, real, and workable building concept.

- The drafter spends less time drafting but more time sketching to prepare data for computer input. Drafters may become computer operators and do drafting directly at the computer terminal. When that happens, the pace of their technical training and accumulation of construction knowledge accelerates. They gain far more experience in much less time than is possible with traditional, on-the-board pencil pushing. This in turn generates greater problem-solving expertise plus more time to deal with higher-level problem-solving activity.

- The engineer, designer, and drafter do computations far faster than ever before. That allows new time to elaborate on the design and test more options than used to be practical. And when they make a change in the design, whether in structure, piping, or ductwork, the repercussions are recorded throughout the system and all related members are redesigned to meet the changed condition. That opens up even more time for refinement and elaboration of the engineering design process.

- The cost estimator gets construction cost data faster than ever. That opens up time to keep pace with the designer in checking cost implications of new design refinements and elaborations.

- The project manager does more managing in less time. By using techniques like Guidelines Systems Management Worksheets, the manager can checklist all contents of the working drawings. That tells all staff members what needs to be done and later, during checking, tells the manager what's been done and what hasn't. This streamlines the whole supervisory operation while vastly reducing errors, omissions, and lapses in coordination.

ANOTHER ADVANCE: SYSTEMS MANAGEMENT WORKSHEETS FOR PROJECT MANAGEMENT

The members of the project manager–designer-client team will gain extraordinary productive power with another tool: Guidelines Project Management and Documentation Worksheets. Each time you check off a building component to go into a job, that act will simultaneously and automatically call up the following information:

Working drawing notation for all related drawings

Relevant standard keynotes and assembly notation

Standard details to use and their locations in file

Outline specifications

Building code and other regulatory information

Technical reference information

Sources of related product information

Names, addresses, etc., of the suppliers of any named products

Consultant and trades coordination data

Construction costs

All this information and more becomes identified all at once in less than a second as each decision is made or tentatively made.

At the same time, every decision and its reason will be dated and recorded, as will changes and their reasons and effects during the job. The end result is complete, day-by-day, job history documentation.

While early versions of such programs will be passive, later ones will be "interactive." That means the designer, manager, and client will feed essential information about the project into the computer, such as size limitations, budget, and spatial requirements. The computer will then make certain automatic assumptions—as would a person—about the structural system, construction materials, mechanical system, etc., and offer those as questions or suggestions to be accepted, rejected, or modified.

If, for example, the design team names a high-quality standard for products and finishes, lesser-quality items or methods of construction won't be presented among the decision-making options in the computer checklist. This automatic narrowing of options to decide on will further accelerate the already high-speed process.

The interactive management checklist system promises to accelerate and simplify project management as never before. Won't that reduce the job time—and employment—of architectural and engineering project managers? No. Once again, the old work takes less time, and new work can be performed in the time saved. (There's a more detailed description of the total interactive project management system in the last part of this book.) The new work this method makes possible is a total building management service for the clients.

Design firms will store virtually every piece of data that can exist about a project and keep it all organized and accessible. Then project managers and clients will be able to predict future design, documentation, and construction time and costs more accurately. They will be able to locate any piece of furniture or equipment and call up reports on every aspect of maintenance, equipment, energy use, tenant spatial allocations, utilization factors, cash flow projections, construction detailing, etc., in almost endless scope.

INTEGRATING COMPUTER GRAPHICS WITH REPROGRAPHICS

Many specifics that follow in this section are subject to rapid obsolescence because the CADD industry is in such explosive flux and growth. But the principles involved will not change for a long time. You'll find them extremely useful in making best use of a CADD system whether it is in house or part of a service bureau.

The first point is this: Advanced systems graphics does not stop with a "final" design on a CRT screen, nor with the printout or plotted drawing. As far as many aspects of reprographics go, that point is only the beginning. When a drawing comes off the plotter or the electrostatic printer, or out of the COM[1] machine, it then has to be reproduced either as progress prints, bid or construction prints, or final as-built record prints.

The reprographic end of the whole process is not well-understood by people on the computer end. This lapse is creating endless pointless misunderstandings. Why, says the service bureau, would you want pin-registrated base sheets and overlays when they can print base and overlay images all together on one sheet for you?

The reason, you explain, is that you want to diazo-print the final bid set with the base screened and the overlays in solid line.

What do you want that for? They can print combined base and overlay images with the base image drawn "lightened" so that it'll be subdued in printing.

You explain further that you may do some final printing in offset at reduced size. For offset printing, you need negatives made that show only base image on one sheet and overlay images on another. For that you have to give the printer base and overlay sheets, and they have to be in perfect registration.

---

[1] COM means "computer output to microfilm" and refers to burning an image into film rather than drawing or copying it. It has great potential value in creating negatives for photo-blowback reprographics. The negatives would be created directly by controlled light beam instead of through straight photography.

What do you want that for? They can print final prints for you already in colored ink and at any scale you want.

But offset is different. An original in color makes a nice show piece, but you can't reproduce it in quantity via offset or diazo printing.

Or you might explain that if they're still using a pen-and-ink plotter, and it takes 30 to 40 minutes to plot out a complete floor plan E-size drawing, they can cut total time in half and reduce your bill by *not* printing the floor plan background with every item of variable overlay data. All they have to give you is *one* floor plan base sheet and a half dozen sheets showing *only* overlay data on each such as plumbing, HVAC, reflected ceiling plan, lighting, electric power, framing, variable partitioning, and furnishings. You relieve the plotter from having to redraw the floor plan in combination with each of those layers of information. Then it's a matter of going to registration on a vacuum frame or to photography to create final reproducible prints from which to print in volume. It may take an $80-an-hour pen plotter 15 minutes to draw the floor plan as background on every sheet. It takes an $8-an-hour graphic work assistant five minutes to achieve the same final result on the vacuum frame.

Some computer experts will argue against all this strictly out of lack of perception of the entire reprographic process that architects and engineers have to go through to produce job prints. Or they may argue it correctly at some time in the future because they do in fact have a new answer that supersedes all the foregoing. Watch for it either way.

At the moment of this writing, mid 1982, electrostatic printers promise to speed printing time enormously so we can get a good reproducible drawing in a matter of seconds instead of half an hour. But it's still all promise. Quality control is a problem, and the machines are prohibitively expensive for most design firms, $80,000 or so. Later, of course, all this will change.

But changing technology does not change the need for ongoing integration of manual, reprographic, and CADD systems. For example when a job comes off the computer and there are still changes to be made, it's more sensible to do the changes on the final output and not go back into the machine. Those who face this situation currently make use of base and overlay or paste-up techniques or both to speed the process.

There will always be times when the machine is plain full up and you can't get in for some real fast input-output. In fact that will become the rule for many high-powered systems. They will always have a backlog of work in the pipeline to ensure maximum utilization of the equipment. That means either you get in line and wait, or you sometimes bypass it all and head for the light table and vacuum frame.

Some small-office computer users already have a good system of interchange between reprographics and their computer. They can't afford a larger plotter, but they can do simple graphics on the screen and produce reasonably good line drawings on a small table top plotter. So they design repetitive elements on the screen, have them drawn individually on the plotter, then make sepia film or photocopies of the elements for manual paste-up and vacuum frame reproduction.

That's one interim approach to maximizing the value of each technology—putting each to best use within its limits and then augmenting it with others. As for the later future, systems analysts are looking forward to strong use of telecommunications, video disks, and the diminution of the mass printing process. That is, construction documents will be delivered in the form in which they're created—electronically from computer to computer.

Recently I was privileged to be on a panel of "systems consumers" sponsored by a marketing company researching CADD and ink plotters and the like for some unnamed company. One participant, Pat Schilling, vice president of Oakland's CADD service bureau Design Logic, created some static with the researcher by insisting that there was no real reason to be concerned with tracing paper versus polyester or graphite versus ink since all future documentation would just be on tapes anyhow.

Later I was asked about my views on the future of all this, and I suggested that future documents would be multicolor three-dimensional (3-D) video images on flat sheets we could roll up and stick in our pockets. And I mentioned the 3-D, full-size hologram system of creating designs and finished structures described in the last chapter of *Systems Drafting*. I also mentioned some other future-oriented aspects of design and construction documentation for orbital and outer space work, biological architecture, and other similar projects. The marketing researcher got a little anxious about this and cut in to ask what others on the panel thought of all this. He, as I, was surprised to hear a chorus of "Oh, yeah, that's definitely coming." and "That's very nicely put; just what I've been thinking about." And "Sure, no doubt about it and furthermore. . . ." And so on. I said to myself, "I always thought there was something great about these systems nuts."

CAD stands for "computer-aided design" and refers mainly to systems for drawing three-dimensional objects on a computer terminal screen. CAD programs are most widely used in manufacturing and industrial design.

CADD refers to "computer-aided design and drafting" and is more oriented toward architecture and engineering. (CAM means "computer-aided manufacture" and often refers to a process of hooking up the results of CAD directly to automated manufacturing machinery. Thus you'll often see or hear reference to CAD-CAM. Most of the architectural and engineering computer graphics in use today derives from early research in CAD-CAM.)

CAE refers to "computer-aided engineering" and may include both computational programs and graphics.

AD means "automated drafting" or computerized drafting. That refers to the process of storing graphic data, manipulating it at a computer terminal, and then having it drawn automatically by a high-speed ink or electrostatic plotter.

INTERACTIVE GRAPHICS refers to drawing with a computer, with the computer acting as a "partner" or prompter that asks questions and otherwise prods and assists the design process.

DATA BASE refers to stored data—written, graphic, or mathematical—that can be reused or kept close at hand for ongoing revision and update.

Fig. 10-1. Glossary of common computer terms.

Fig. 10-2. Computer terminal with CRT for inputting and manipulating drawing data.

(Courtesy of Everett I. Brown Co.)

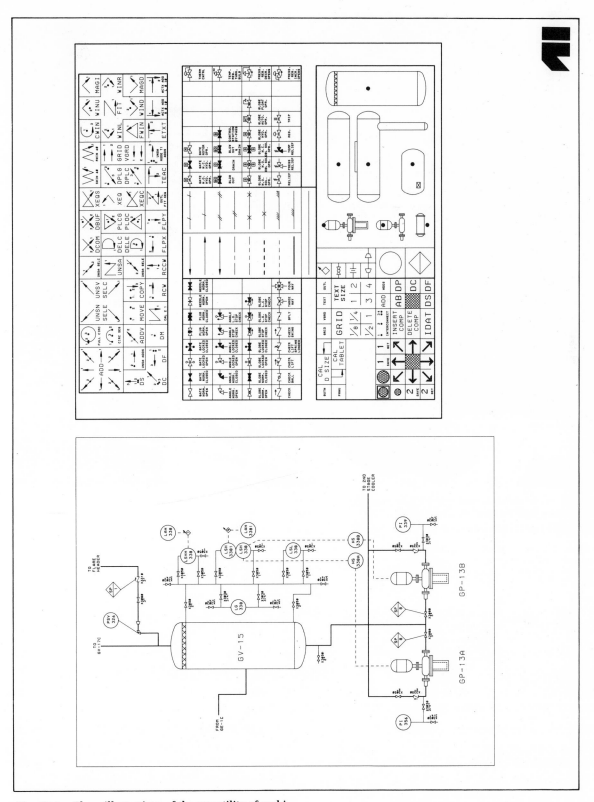

Fig. 10-3. Three illustrations of the versatility of architectural and engineering CADD. First a mechanical schematic drawing shown with the "menu" of pieces and components that can be assembled on the video display and later drawn automatically. *(Illustration turned on page.)*

(Courtesy of Applicon.)

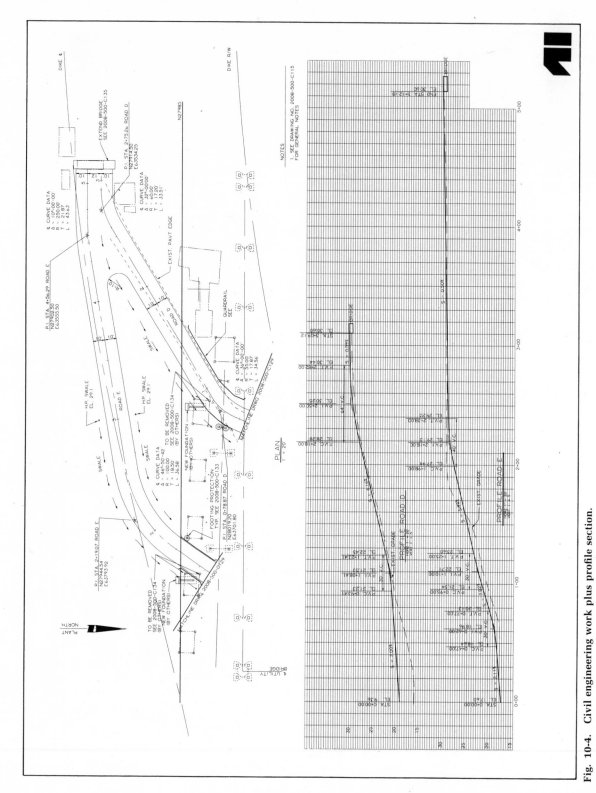

Fig. 10-4. Civil engineering work plus profile section.
(Courtesy of Applicon.)

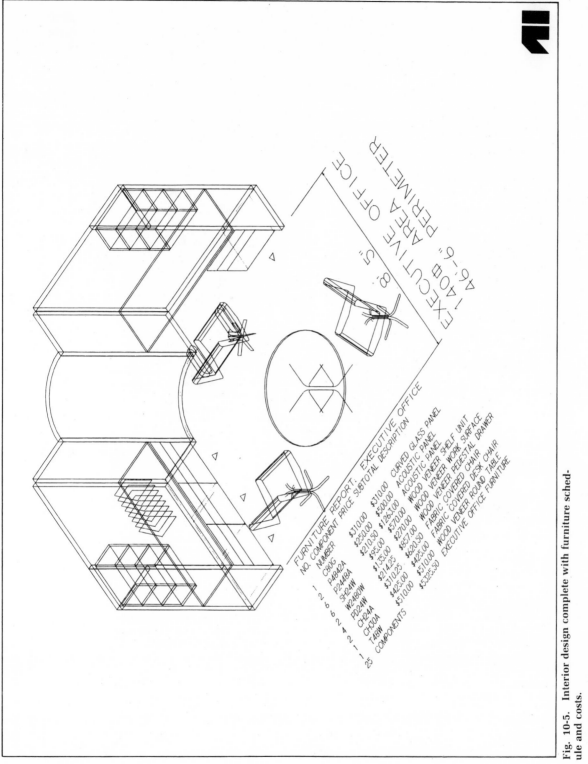

Fig. 10-5. Interior design complete with furniture schedule and costs. (Courtesy of Applicon.)

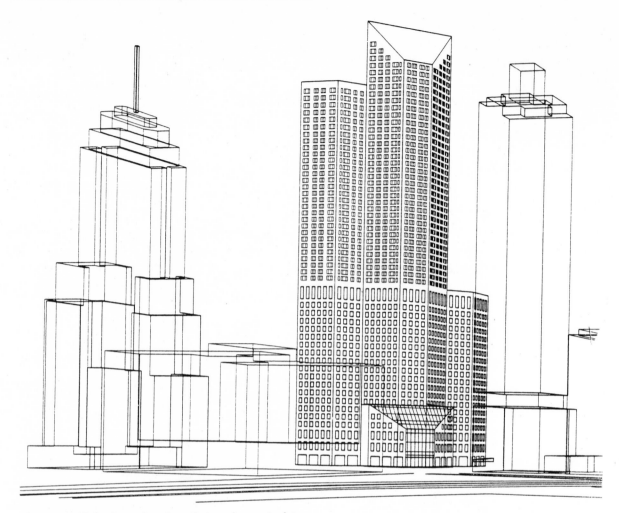

Fig. 10-6. Skeletal exterior views that can be manipulated for design studies, shade and shadow studies, structural deformations, etc.

(Courtesy of Skidmore, Owings & Merrill, Chicago.)

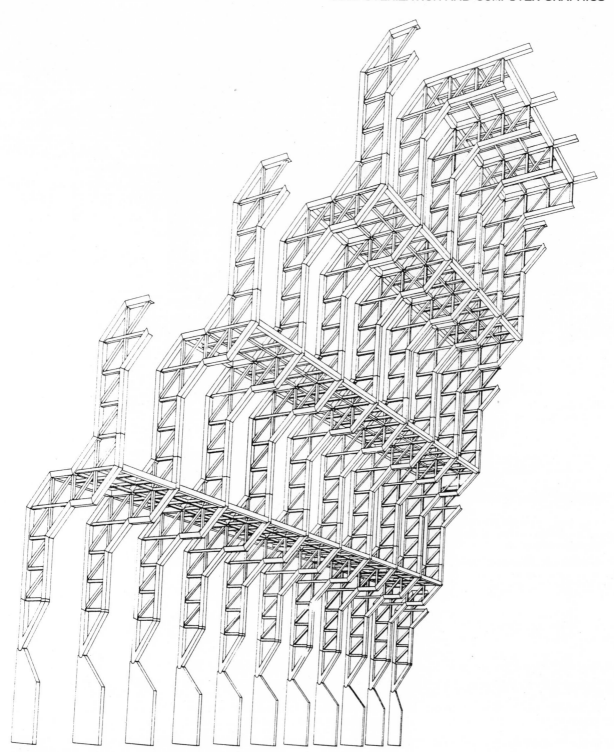

Fig. 10-7. Structural sections used in design studies by
Chicago SOM.

(Courtesy of Skidmore, Owings & Merrill, Chicago.)

11

CHECKLIST OF STEPS IN RESEARCHING COMPUTERS

USING THE CHECKLIST

This is a model checklist of actions for you to take when researching and shopping for computer systems. Add your own ideas to the checklist. Skip those that don't apply to you, and create your own customized plan of action. Use the checklist collectively as an agenda for decision-making meetings and as a guide for delegating.

This checklist is specifically tailored to the needs of architects and engineers. Those needs are much more complex and quite different from the needs of most business users of computers. As you'll see, these differences are hard for many computer consultants and computer company representatives and consultants to grasp.

The checklist is thoroughly adaptable for researching and planning the purchase or leasing of equipment or systems of any size or type, from personal computers to mainframes.

The steps checklisted below involve four main phases of activity required to research the system(s) suited to your office. Please note that steps 2 and 3 are in one sequence during research but are reversed as you specify and acquire your system.

In specifying and ordering equipment, you must know needs first, what software and storage will handle those needs, and then what hardware or turnkey system will handle the

software and storage. But to identify your needs accurately, you also have to find out what's available in equipment and software. Otherwise you can make a totally unrealistic needs list. So do research on equipment, systems, services, and software to see what's available, then build a needs list, and finally backtrack again from needs to software to hardware and services. In other words, switch back and forth between phases 2 and 3 listed below as you gather your data.

The main phases of research and action are:

1. **Gather resources.** We get calls and letters every week from architects and engineers who have made uninformed, bad, costly choices in computer systems. They take shortcuts which make them dependent on and vulnerable to sales representatives and unqualified consultants. Gather together as much pertinent information as possible and read it. The steps listed below deal with where and how to arm yourself with reliable data and how to tell the difference between good and "not so good" data.

2. **Identify what you need.** Many architectural and engineering firms buy equipment, plug it in, and hope it will do something. Then they learn, gradually, that they have a lot of ground to cover to find out if they've made the right choice or not. A sensible way to shop for computers is to reverse the usual steps in the process. Shop for hardware as the last, not the first, step as follows: (1)

Identify what you need done that a computer can economically do. (2) Identify the software and computer storage capacity that can handle what you need done. (3) Identify the equipment or services that can handle that software.

3. **Identify the best equipment and services.** When you know what you want done, compare the software, computer memory capacity, number of workstations, etc., with likely income you'll generate, money you might save, and what you can realistically afford. From that evaluation and comparison you'll make a "short list" of what to look at more closely. You may find, for example, that no one makes exactly what you need as a ready-made package, so you'll have to have one created for you. Or you may learn that the best choice for you right now is not to choose.

4. **Prepare your staff and management.** The right software and hardware still don't add up to a functioning system. People problems—a combination of misinformation and fear—can destroy any system, anytime, anywhere. You'll need to implement a combination of education and proselytizing to ease the office through tricky and politically treacherous changes.

We've just outlined the broad picture of research and preparation; a more detailed consideration follows.

CHECKLIST OF STEPS IN COMPUTER RESEARCH

GATHERING RESOURCES

_____ Acquire your library of required reading and reference data immediately. Prioritize and schedule reading of the most recommended texts. See the _Guidelines Reprographic and Computer Resources Guide,_ cited at the end of this book, as a source of references.

_____ Mark out a time frame for implementing the research, specifying, office planning, etc. Identify the hours of the day that can go to the project on a regular basis. Budget specific amounts of time on a weekly basis for reading, equipment demonstrations, field trips, etc.

_____ Locate architectural and engineering computer user groups in your city or region, and ask around about other firms similar to yours that have moved ahead in computerization.

_____ As part of the preceding step, start a notebook directory of the people and resources you come across that may be useful to you later.

IDENTIFYING YOUR COMPUTER NEEDS

_____ Begin a starter list of desired functions and uses of computer systems in your firm. Add a tentative list of "weights" of 1 to 10 for their relative importance.

_____ List possible long-range future uses and give them weights according to likely importance.

_____ Identify the likeliest first uses. Give them weights.

_____ While reviewing your needs and steps in researching, weigh the pros and cons of hiring a consultant to help in systems research and selection.

_____ Make general assumptions on whether you

____ want, can use, and can afford a personal computer (PC), microcomputer, minicomputer, mainframe, or any combination of these.

____ Begin a tentative list of criteria that will bear on your decision: location of manufacturer, proximity of service facilities, etc. Give these weights of value.

____ Do a starter cost-benefit analysis.

____ Start a review of your facilities planning and look for radical and expensive changes that may have to accompany computerization.

____ Obtain product literature from all likely sources.

____ Contact sales representatives from the strongest-looking companies that serve your area and ask for a list of *all* architects and engineers in your vicinity who have their equipment.

____ Contact architects and engineers in your area who have equipment you're interested in. Knowing that some will be happy to share all the data they have and others won't, pursue as many questions as possible.

____ Contact architectural and engineering computer service bureaus in your area. Ask for literature on their equipment, services, and prices. Ask for the names of *all* architectural and engineering firms that have worked with them.

____ Contact architectural and engineering users of computer service bureaus in your area and do a line of questioning like that suggested for buyers of systems. Keep in mind that many problems architects and engineers have had with service bureaus are due to their own inability to organize their work properly for inputting.

PREPARING YOUR STAFF

____ Assign a staff or staff-management task force to conduct initial screening of systems. They're not likely to go far wrong, and it will early on give them valuable familiarization with systems technology.

____ Identify the most "computer-wise" staff members in your firm and include them in meetings, general planning for staffwide education, etc. They will help act as conduits of information to others in the office, and they will be likely candidates for employment in your expanded computer operations. (Warning: Do not allow the computer enthusiasts to form a club among themselves that excludes others. Their role is to help gather and transmit information, not become possessive guardians of it.)

____ Arrange lunch time and Saturday field trips to architectural and engineering computer service bureaus in your cities and to the computer facilities of other architectural and engineering offices.

____ Establish a budget and timetable for sending staff and management to architectural and engineering computer short courses, trade shows, workshops, etc. Watch for notices in the architecture and engineering computer-oriented periodicals and newsletters.

12

COMPUTERIZATION: THE EQUIPMENT

THE HARDWARE THAT DOES SO MUCH WORK

The usual computers you see advertised these days are the "microcomputers." Microcomputers are roughly divided between the "home," or "personal," computer, costing from $300 to $5000, and the "business," or "desk top," computer, costing from $6000 to $20,000 and up, depending on peripheral equipment. (These numbers represent early 1983 dollars.)

More powerful computers that sell for roughly $20,000 to $500,000 are called "minicomputers." And the "large," or "mainframe," units, as used by government agencies and large corporations, are in the million and multimillion dollar range.

The lowest-cost micros won't do much for any business or professional office except serve as a plaything. The desk top computers, however, are the most versatile and helpful office time- and money savers to come along in years. Small firms find them to be a boon, and, while larger companies use bigger systems for in-house work, it's common that management and employees will have desk top micros for their personal use.

For any serious graphics or drafting work, especially for multiple terminals, you'll have to use a minicomputer. You won't be concerned with a large mainframe unit unless you use time-share services or service bureaus that have such equipment.

A SHORT COURSE ON COMPUTER OPERATIONS AND VOCABULARY

The first and most fundamental point about the machines is that they don't do the things most people think they do. They're just tools with which you store information and move it around. They'll perform certain functions for you and move data around according to the patterns and rules that you establish, but they won't do any of it on their own. Architects and engineers who acquire the equipment often are not aware of this elementary limitation. They expect the computer to have motivation and a memory of its own and to go to work the moment it's plugged in. They plug it in and spend months afterward trying to figure out how to make it do something.

Although long called "thinking machines," computers think or know absolutely nothing. They just provide microscopic switches and circuits for low-level electric current. "On" means "yes" or "one" in computer language, and "off" means "no" or "not one." With that elemental yes-no starting point, you can build any possible body of data.

Each single on-off circuit switch is called a "bit." It takes a combination of 8 bits (this number may vary slightly) to store a single character such as a letter or number. Such an 8-bit character, or symbol, is called a "byte." **119**

The capacity of computer storage or memory is normally measured in terms of how many bytes it can handle and is expressed as so many K, or KB, for kilobyte, which means 1000 bytes. For example, a 32K memory has room for 32,000 bytes, or characters, before it's filled up. (These numbers are close approximations for simplification. In actuality, a K or kilobyte represents 1024 bytes. A megabyte, or MB, means 1 million bytes.)

The "brains" of the computer—its circuitry—are the primary part of what's called the "hardware." It's all potential and can't do a thing without instructions. The instructions are what is called "software." This provides the precise step-by-step operating sequence for storing, manipulating, and retrieving the data in the machine. Software will cost you more in the long run than the equipment—a point to budget for when you are planning to install and operate a system.

INPUT EQUIPMENT

Besides the circuit boards that comprise the core of the hardware, other hardware is for either input or output. The dominant input device these days is the keyboard. It is very much like a typewriter keyboard with additional command keys for special operations. You can also input data with the old-type computer cards, but the cards, and the keypunching operations that go with them, are mainly obsolete.

An input keyboard is usually connected to a terminal display. This is the cathode ray tube (CRT), much like a television screen and the sort of thing you see everywhere these days, even on personal home computers. What you type on the keyboard is shown on the screen.

Another input device is called the "digitizer." It's a surface with a built-in wired grid that lets you input or record positions in a two-dimensional plane: lines, points, symbols, etc., all by coordinates. The "memo," or "tablet," is a similar device with which you can call up stored symbols and graphic data by touching marked parts of the tablet with a stylus.

Some input is possible with the optical scanner. This is a device that "reads" existing data such as typewritten text and records it automatically in the computer file.

OUTPUT EQUIPMENT

Output hardware includes printers for letters and numbers (alphanumeric) and plotters. Printers include "dot matrix" printers, which are relatively inexpensive devices that print letters and numbers in dots. You've probably seen these in letters from any of your friends or relatives who are home computer enthusiasts. There are "line printers" which type full lines of data in one shot—fast but low in print quality. There are high quality impact printers which use an interchangeable type head called a "daisy wheel." These give you the best-quality printing currently available. A strong runner-up is the laser-type electrostatic printer that combines computer imagery with photocopy technology. It is the fastest of all and, once quality is up, will be the standard of the industry.

The common output device for graphics is the "plotter," an automated drafting machine that draws in ink on either paper or polyester drafting media. The likeliest replacement for the old-style pen-and-ink plotter is the electrostatic plotter. New ones combine computer, fiber optic, and photocopying technology to produce line drawings much as an office copier would—and just as quickly. Still another speedup option with great promise is computer output to microfilm (COM). Instead of drawing with pen and ink on drafting media, the line images are etched on photographic film. This has good potential for those who will marry their computer graphics with overlay reprographics.

INPUT-OUTPUT DEVICES

The CRT terminal screen is also usable in some cases as both an input and an output device. That is, you can manipulate data directly on the screen with a finger or stylus, and there are ways of photographically transferring information directly off the screen onto print paper or film.

Another combination input-output device is the telephone coupler, or "modem." It allows direct telephone transmission between your computer and others. That's how you'll use time-sharing services to hook into more powerful machines on a time charge basis. It's also how you'll take advantage of the many software and data exchange networks that are springing up nationally.

Although computer circuitry has its own storage capability, most software and data storage is recorded either on tape or on hard or soft "disks." Disks are either rigid or "floppy," like ultrathin versions of old 78 or 45 rpm phonograph records. The built-in memory of one popular model of desk top computer is 48K (approximately 48,000 bytes, or characters). One small diskette can store 300,000 bytes. You can buy and use as many disks as you want, so there's no real limit to your possible memory storage file. Disks are common for text storage, but tape is considered more practical for the large amount of data involved with CADD.

Each computer also has an "operating system," which is built-in software that instructs the machine how to handle the data it gets. The operating system takes up memory space. And any time you set up a disk or tape to record data, it will also store instructions—its programs or software—for telling the printer what to do, keeping track of what's filed in the machine and how much memory has been taken up, etc. All these instructions take up space in the computer's memory and on the disks and tapes. So when you see the plain memory capacity expressed in bytes or kilobytes, keep in mind that the actual memory available for inputting new information may be far less.

The popular disks for most personal or microcomputers are 5¼ inches in diameter and are usually called "diskettes." They come with varying storage capabilities, and some can store an amazing amount of information. Many microcomputers use a larger disk size, 8 inches in diameter. You also have the option of storing data on hard disks, which are considered permanent memories.

Disks and diskettes are handled in what are called "disk drives," which can be likened to a record player. After turning your machine on, you place whatever disk you want to work with in the disk drive. The drive spins the disk from position to position, and a "read-write" head, rather similar in principle to the pickup head in a tape recorder, transmits data from the computer's internal memory to the disk and back again.

Data on disks or tapes are not recorded and placed at random. They are stored in storage spaces, or "files." When you begin work with a computer, you will create a file space or call up a file that's already been created according to its name or number. The file stored on disk or tape is then copied and recreated in a "buffer memory" inside the machine. So the work you do inputting or manipulating data isn't done on the disk, it's done in the computer's memory.

Later, when you want to record what's been done or move it to another file, you return to the storage disk, quit the file, and move on to the next. It's good policy to "write out" on disk or tape whatever new work you've created periodically—even every half hour, and certainly every time you leave the terminal for any period of time. The reason is that if there's a hardware failure or power failure, or if someone starts working at your terminal, you risk losing the information you've just put into the machine memory. This happens frequently early on in computer use.

You have to write data into the memory of a disk or tape for permanent storage. On top of that, you'll want to make periodic, even daily, backup files. These are separate copies of data on disk or tape. They are needed because disks or tapes get physically misplaced sometimes. They can be erased by accidental proximity to magnets or certain electric equipment, or they can be damaged by hardware failure. They can also be erased by accidental or deliberate command.

A final note on software: One of the most common questions about computerization is, "Do you have to learn programming?" The answer is no. If you make a major investment in equipment, you'll want someone around who can write special software and modify software you purchase. But for most work you'll be doing, you'll buy and use the software ready-made.

Fig. 12-1.   A CADD workstation. The terminal screen on the left provides a directory of commands and prompting data; the terminal screen on the right displays the graphics.

(Photo by Carol Hickler, courtesy of Design Logic, Oakland.)

Fig. 12-2.   The alphanumeric input device, a standard typewriterlike keyboard.

(Photo by Carol Hickler, courtesy of Design Logic, Oakland.)

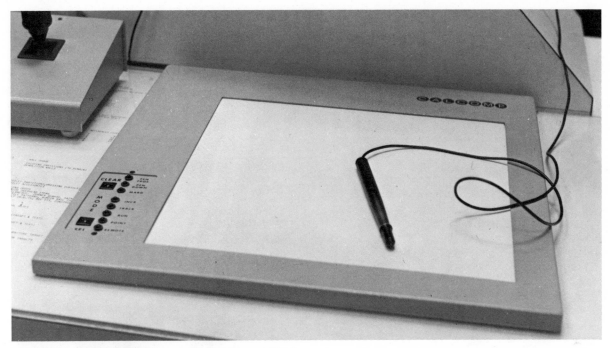

Fig. 12-3. The graphic input device, a digitizing "tablet."
As you mark points with the stylus on the tablet, you create graphic images on the video display screen.

(Photo by Carol Hickler, courtesy of Design Logic, Oakland.)

Fig. 12-4. A graphic control device, a "joy stick" which
controls enlargement and reduction, and movement of the
screen image back and forth.

(Photo by Carol Hickler, courtesy of Design Logic, Oakland.)

Fig. 12-5.   The plotter. Some plotters are positioned as a flat bed, some are vertical like this one; some print by electrostatic technology.

(Courtesy of Everett I. Brown Co.)

13

IDENTIFYING YOUR COMPUTER NEEDS

HOW NOT TO DO IT

It's common in computer shopping that people become very excited after seeing demonstrations of equipment. Then they push numbers around to figure out how they can afford it. Once they persuade themselves they can afford it, that ability becomes their primary justification for going ahead with it. Before long they come to learn that their motivation was misplaced and their criteria for selection nonexistent. It's a form of impulse buying—on a grand scale.

At this point it's no exaggeration to say that most computer purchases are seriously flawed in major ways because of inadequate analysis of need. Here are some true-life case studies that speak for themselves.

CASE ONE: HOW ORIGINAL INTENTIONS GET LOST IN THE SHUFFLE

A multidiscipline firm was prodded into shopping for a computer to do architectural and engineering working drawings. The instigator was their production manager, who had done some extremely high-speed and profitable projects with systems drafting. He wanted to build on their advanced systems by adding computer-aided design and drafting (CADD) capability.

Top management agreed to look into computerization, but no one in the firm knew exactly where to start on it or what to do. Someone suggested hiring a consultant. Nobody knew exactly who would be qualified so they contacted a local high-ranking architectural school. A professor recommended someone from a computer research group associated with a prestigious eastern university.

The consultant came to visit and spent a week interviewing staff and management. His report was presented to the partners of the firm, and they voted to purchase a well-known, high-powered minicomputer and to establish a computer center.

The production manager who initiated the proposal was one of many who participated in the final decision. His recommendations were virtually ignored by the consultant. The consultant, like most academics, had no knowledge or interest in working drawings and computerized drafting but did specify enough "space" in the mini to handle such work.

The firm now has a powerful minicomputer to handle accounting, word processing, and other work that could be done by a much less powerful micro. There's no ready-made software available to create a CADD system for that equipment. The consultant took it for granted that the office could write software to input floor plans and elevations but didn't conceive they would really want to create complete working drawings. "Why?" he asked the production manager sometime later. "Why would anyone want to do that when it would be so much cheaper just to have some drafter draw them?"

CASE TWO: BEING TAKEN BY THE DEMO

A firm of housing designers with a staff of 20 purchased a turnkey drafting system for $200,000: three workstations, a plotter, check print plotter, digitizer, and packaged software for drafting.

Two staff members were trained to manage the system several months in advance of installation. Three months after installation, friction developed, and the first computer management team quit the firm.

A second team was selected from senior drafting staff and trained to run the system. Nine months after installation, the equipment was helping drafters produce drawings at a cost and time ratio of 3 to 1. That is, it was taking three times as long and costing nearly three times as much as normal hand drafting to produce drawings. Mainly, it was taking far longer to input and record design and drawing changes than they ever imagined it would.

The problem started with the shopping. The demonstrations that seemed promising were done on a single-terminal workstation. The office set up a three-station system using the same computer power, thus slowing down response time at all the stations.

The demonstrations showed packaged situations and did not reveal deficiencies in the software. The demonstrations showed it was easy to move lines and symbols around the terminal screen. That is what impressed the buyers. But the demonstrations didn't show some peculiar problems of their diskette storage system, or the plodding pace of inputting graphic data. The firm's principals assumed that most systems were essentially alike and didn't see the point in spending time and money on detailed research. They relied on comparison charts provided by the sales representative.

The firm now has an underpowered system. They've committed or spent $300,000 in nine months and still can't get usable drawings out of it. A consultant called in to help out doesn't envision the system being usable for another six months. Even at that point the system may only approach a 1 to 1 efficiency ratio compared with hand drafting.

CASE THREE: THE DANGER OF BROAD ASSUMPTIONS

The operations manager of a large (200 staff) federal government architectural and engineering department argued against systems drafting, preferring, instead, computerization. He was an electrical engineer and had correctly perceived significant efficiency gains through computer drafting in electrical engineering. His intent was to input plans of all the existing buildings the agency was responsible for in the machine. With this graphic data base they could just pull plans from the computer file and redesign them for remodeling directly on the computer terminals.

The agency hired a consultant and software writer. The consultant had created a building-plan inputting system for use on fairly low-priced ($100,000) equipment. It would also do some modest 3-D line drawing and could store furniture and equipment lists. The consultant had no architectural or engineering drafting experience or knowledge and often answered staff questions about his system's capabilities by saying, "Why would anyone want to do that?" or "We can't do that yet, but it should be possible in a few more months." Though hired as an independent consultant, the individual actually acted solely as an agent and sales representative for his system, his software, and his set of equipment.

The equipment was purchased and put to work. It had only one terminal, and inputting floor plans went very slowly. It became apparent that by the time most buildings in the domain were recorded, they would already have been changed. In addition, the drawings used to input data on existing buildings were mainly obsolete. So someone would have to go out and measure the existing buildings to input the most up-to-date data. With two people hired just to take measurements, it was estimated that it would take several years, and the project was dropped.

The equipment gradually went out of service, and a new, more powerful system was ordered. The main function of the new system was, as before, to enter existing buildings and record changes. It had the power to do that, but the agency was still stuck with the dilemma of inputting newly created measurement drawings.

And, every time a minor change was inputted for drawing in memory, it took several minutes for the computer to go over the content of an entire drawing just to record even the simplest change. The new system has proved too slow for changes. Since that was one of its two main functions, it is being phased out of service and the operations manager is shopping for a larger system.

CASES FOUR, FIVE, ETC.

A small firm bought a four-station unit and discovered more uses than they expected. Unfortunately response time slowed down so much when all stations were in use that one terminal had to be taken out of service.

Another small firm bought a single-station unit advertised as suited to specification storage, financial management, and other architectural and engineering uses. The problem is that it can only be used by those who can elbow others off the terminal. Work is put aside until the computer is available. It's become a bottleneck in the work flow.

An engineering firm bought several popular-brand microcomputers as an aid in structural calculations. Staff members now point out that virtually all the computation software can be matched in speed and accuracy by the programmable calculators they were already using.

What's common in all these cases? No systematic analysis of need, no realistic understanding of machine capabilities.

Here's an outline of steps on how to go about establishing your needs:

1. List all the functions you're interested in. As a guide, look at the architectural and engineering uses described in Chapter 10. Keep in mind that you may not want one system to do all the things you want; you might want multiple systems. You might want some work done by service bureaus, some by micros, some by a mini. All of this is clarified as you proceed through the research process. As noted in Chapter 11, give some numerical value, or "weight," to the importance of each function to the office.

2. Judge what you believe to be the earliest, best uses of a computer system from the list cited above. Where would you most sensibly like to start?

3. Begin analysis of work that's done now without computers, and compare time and cost with potential computer efficiency.

4. Review work that you don't do now but could do with computers, and estimate, in general terms, the value of that work.

5. List needs you have that are hard to quantify. "Intangibles" should be included in every analysis of computer needs, costs, and benefits. Add those to your list and give them weights too.

6. Once you have the larger picture of your needs, proceed to compare your tentative conclusions and options with what's available in terms of software, hardware, and service bureaus.

Here are conclusions commonly reached by firms that pursue methodical comparisons of needs, software, hardware, and costs:

Most small design firms conclude they want immediate, measurable benefits in efficiency and earnings. That requires that they put ambitions for a CADD system aside in the first go-around and focus on an economical workhorse that'll handle financial management, basic engineering, and word processing. That usually adds up to a multistation, expandable microcomputer, or a few stand-alone units.

Some small to medium-sized firms want long-term, powerful potential graphic and drafting capability, but want to start small. The options are to buy either the lower-priced complete systems or a single station, also low-priced but with hefty add-on potential. All other things being equal, that second option is working well for some smaller to medium-sized firms that have strong growth prospects.

In a few significant cases, individuals want the power of high-priced equipment just for their personal use. They don't want to manage or supervise staff. They just want to design and do drawings and specifications in the fastest and most enjoyable way possible. (They're like the highly productive individuals who have armed themselves with a vacuum frame and other graphic work center equipment and do the work of four to eight junior drafting staff without all the headaches of staff supervision.) This is an exception to the rule that computers don't replace people. In this case they're specifically brought in for that purpose because of personal preference. The owner computes the hourly cost of drafting staff and then estimates whether she or he and the new machine working together without staff will cost and earn the same amount as the staff. If it balances out, the owner lays off unwanted staff and makes a new start with the machine.

PAYING FOR COMPUTERIZATION

One useful way of considering the cost and return on an investment in equipment is to compare it with the cost of people. The idea of investing many thousands, or even hundreds of thousands, of dollars in equipment makes most architects and engineers nervous. Not that they don't spend such sums; it's just that the money is spent over a span of time, as in weekly wages, and not thought of as lump sums.

If you pay a drafter $8 per hour and have costs of another third of that for insurance, holidays, sick pay, etc., that's about $10.65 an hour. This times 40 hours per week equal $426. Over the course of 52 weeks it equals $22,152 per year, and in five years equals $110,760. That's just straight pay with normal perks.

Naturally no one charges clients at straight wage rates. That wouldn't cover the numerous overhead expenses required for each employee nor allow any percentage of profit margin for the firm's future growth. Firms that charge clients on an hourly basis normally charge a multiple of between 2½ to 3 times direct personnel expense to cover overhead and profit ("direct personnel expense" equals salary plus benefits). That affects the comparison between equipment (or outside services) and staff considerably. In the case above, a 2½ times multiplier raises the hourly dollar amount to $26.63. The weekly rate (at 40 hours) is $1065. The annual rate becomes $55,380 (assuming no time off). And five years' worth of wages is $276,900. That amount is close to the price of a large machine. It's close, that is, if you're paying cash for the equipment and if you ignore all the other costs. Those financing and support costs will more than double your final real expenditures. However, a large part of the total costs can be recaptured through tax credits, depreciation, and salvage value.

To illustrate part of the cost-and-benefits analysis process, let's see what happens in costing out a microcomputer.

Suppose you manage a small office of eight employees and three partners, including yourself—one for design, one for business, and one for production. There are six staff on the drafting boards and two support staff.

You've done your equipment research and favor a two-station micro. One terminal is for the business manager partner to do word processing marketing chores, feasibility studies, and miscellaneous repetitive office paperwork. The other terminal is for the production partner for specifications writing, engineering computations, door and finish schedules, job scheduling, and a keynote system. The design partner doesn't want any part of it.

You find a well-recommended business desk top micro system with four-disk drive, a good-quality daisy wheel printer, and two terminals with add-on capability for more. Other possible future peripherals include a hard disk, small plotter—the usual array. The best negotiable price is $12,000.

You estimate a three-year practical payback period. By analyzing the amount of repetitive work that can be replaced by the machine, it looks like one terminal will be worth 20 percent of the business partner's time, mainly for overhead tasks. The partner earns $24 per hour including perks and works a 50-hour week. The equipment, if it augments work about 20 percent, is then worth $240 per week for that one station.

The specification writer says she can rent good outline-specification software for $600 plus renewal fees for a total cost of $1000 over three years. Her production time analysis shows that working-drawing schedules, keynotes, and job scheduling functions on the computer should save $6000 a year in labor, or the equivalent of $115 per week. The financial accounting is so well handled by an outside firm that there's no need to include that function as yet.

Therefore, the estimated weekly value in document production, real estate analysis, engineering, and word processing totals $355 per week, or $18,460 per year. Because of the learning curve and start-up delays, you don't expect to earn any return on the system for at least four months. But after computing that out, you estimate an additional improvement in productivity over the remaining two years that will more than balance out the start-up loss.

The price is $12,000. Financing that over three years adds about $5400 in interest. Sales tax is $720. That's a total of $18,120, which will be spread over three years. That divides out to $6040 per year.

How costs and savings add up on an annual basis is illustrated on page 130.

Estimated minor software services and additional purchases such as the specifications package	$ 2,300
Maintenance estimated at $1500 per year	1,500
Supplies (print wheels, paper, floppy disks, files)	2,200
Furniture, new shelving, additional wiring and other minor improvements ($2400 total spread over three years)	800
Total annual support costs	$6,800
Equipment (annual average cost)	$6,040
Total costs	$12,840
Estimated gross payback	$18,460
Net annual payback (gross payback minus costs)	$ 5,620

This is clearly an adequate return, and you haven't plugged in investment tax credits, depreciation, or software-development tax credits.

A BARRIER TO COMPUTERIZATION

Now a few words about what may be the worst problem you confront in computerization: people. Not that people in themselves are the problem. The problem arises from the feelings engendered in them by a new and often unwelcome technology. Employees find themselves required to learn a whole new working process. They didn't ask for it. They don't particularly want it. They don't see it as an improvement over what they're used to.

Nevertheless, here it is, a new imposition by management that seems to make their lives more difficult. Even worse, it may appear to take away the individual quality of their work and turn design professionals into factory workers assembling pieces of buildings.

Let's take a special example. Suppose you've been in the profession quite a few years. You're competent and efficient—particularly in drafting. It took years to develop your knowledge and skills, and your salary and job responsibilities reflect those years of experience.

Then along comes systems drafting or computer-aided design and drafting. Although the systems don't entirely do away with drafting, they radically change the nature of the work. Your familiar skills decline in value, and there are now brand-new skills to learn. That's discomforting.

The frustrating thing about learning new skills is that you're now operating like a newcomer— at entry level like some kid out of school. On top of that, your skill in new techniques may not match the skill of newcomers because they don't have long-standing habits to change. All in all, it's a tough and unpleasant situation.

Management must remain aware of all this when facing what seems to be resistance or even sabotage of new systems. It's not just that people are worried about losing their jobs or just resistant to new ideas. Mainly they're resistant to losing their self-esteem. If you're sensitive to that, you'll cut out a lot of the problems.

Here are a few more particulars about likely responses to computerization in your office.

- According to studies by the Booz·Allen & Hamilton management consulting firm, you can expect one-third of professionals and managers to be somewhat resistant to using the computer. About 10 percent will never adjust to it.

- Although the older staff and management may be more outspoken in doubts or suspicion about computers, age is not a decisive factor in discontent with the new system. What is more important is a combination of age and years already spent with the firm. Those who are older and have put in much of their working lives with the office are more prone to have problems than is an older person who is a newcomer.

- The staffer or manager who already knows how to type will adjust far faster than most people who don't type. Typing is part of operating a computer terminal and, for that reason alone, is now becoming part of standard required business and professional skills. Those who are slow and awkward at a terminal keyboard will bring other problems, mainly emotional ones, into the situation.

- Those whose work is most routine will feel vulnerable because they'll quickly catch on that the computer is especially powerful as a workhorse on routine tasks. Those who have grown comfortable in nondemanding routine will have to change job roles or change offices. Some will welcome the change, but those who don't are those who should *not* be assigned any important role with the computer. They can do an immense amount of damage on a computer and hide themselves from blame in the process.

14

CADD SERVICE BUREAUS

HAZARDS AND BENEFITS

CADD has evolved mainly for manufacturing design and big-ticket industrial engineering such as for military contractors and airplane manufacturers. That's why you'll often see the phrase "CAD-CAM" (computer-aided drafting–computer-aided manufacturing). Current software and services oriented toward architecture and engineering are an offshoot of that development, and are still a comparatively undeveloped one at that.

While you can expect up to 20 to 1 efficiency improvement per workstation for certain schematic engineering drawings, few people are doing any better than 1 to 1 for architectural and engineering construction working drawings. The main advantage of CADD in this realm is as a speedup when there's a time crunch. The final cost may come out about the same, but production time and cash flow can be accelerated on a well-managed CADD job.

Service bureaus are recommended as an entry point to computerization. You get your feet wet and learn about the benefits and problems of computer services without investing in your own system.

That's fundamentally a sensible approach, but unfortunately it won't always hold true. As equipment becomes widely available, you'll see a small army of people rush into the architectural and engineering service bureau business. And as jobs fail, expectations aren't met, and bills aren't paid, you'll see most of that same small army of people disappear from the business.

Some newcomers to the architectural and engineering service bureau business are incompetent. Some are fine in their specialties but don't understand the unique needs of architects or of some special branches of engineering. They'll sometimes take the approach that "accounting is accounting" or "a drawing is a drawing" and assume they can serve design professionals with underpowered equipment and inappropriate software.

Overoptimism is another problem. You'll virtually never hear computer owners underestimate what their equipment and programs can do. But it's the underestimate that's most often true.

Expectations are equally unrealistic among the customers. Computers are supposed to be able to do anything, and you'll hear design professionals say with great pride, "Oh, we're doing all that on computer now." The problem is that they don't really know what's being done on the computer, or how it's done, or whether it's done right or cost-effectively. They don't know until they experience some missed deadlines followed by a whopping bill for services misrendered.

SHOPPING FOR A SERVICE BUREAU

To approach a service bureau correctly, it is crucial that you carefully check referrals from other customers. As I've said before, ask for names of *all* customers, not a screened list. Ask for résumés of the people who run the service. When you're satisfied with references and management's credentials, bring in a sample job for a trial run. One drawing or part of a drawing won't do for a test. Reputable service bureau managers say that as of mid 1982, if you have only $5000 or less to spend to test computer drafting services, you won't be doing enough work to evaluate the system properly.

In order to use a computer drafting service bureau, or an in-house system for that matter, give extra time early on toward simplifying your drawings. Set functional drafting standards and cut overdrawing to the bone. Every line and note the computer has to handle is expensive.

Do an extensive mock-up of the entire job. Carry design development to completion and sign-off by the client before going to the computer for drafting. The computer input operators will need clear but bare-bone sketches with ample, accurate, and wholly coordinated dimensions. The ongoing changes so casually pursued in the typical drafting room will eat you alive when you computerize. And clients must be forewarned of the added reimbursable costs if they decide to initiate changes after computer input has started. Typical hourly fees these days, late 1982, are $30 to $50 per hour for word processing and $50 to $80 per hour for consulting and CADD services.

Service bureaus vary radically in capability and experience. The only way to know what they'll do for you is to start shopping and ask for literature and referrals. The market changes too rapidly to list such bureaus in this book, but you'll find an up-to-date listing in our supplementary publication *The Guidelines Reprographic and Computer Resources Guide*, cited at the back of this book.

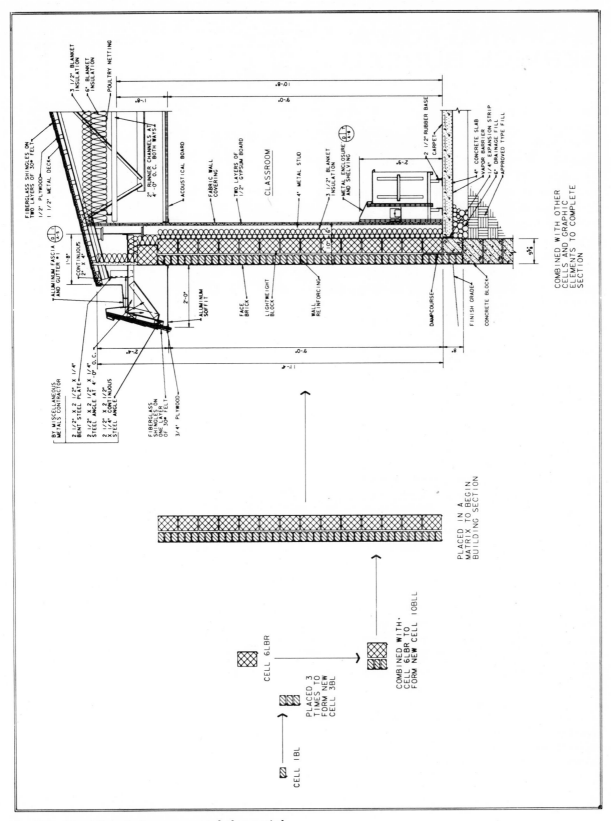

Fig. 14-1. How bits and pieces are created, then copied in multiple and assembled by computer. It's electronic paste-up drafting. (*Illustration turned on page.*)

(Courtesy of Everett I. Brown Co.)

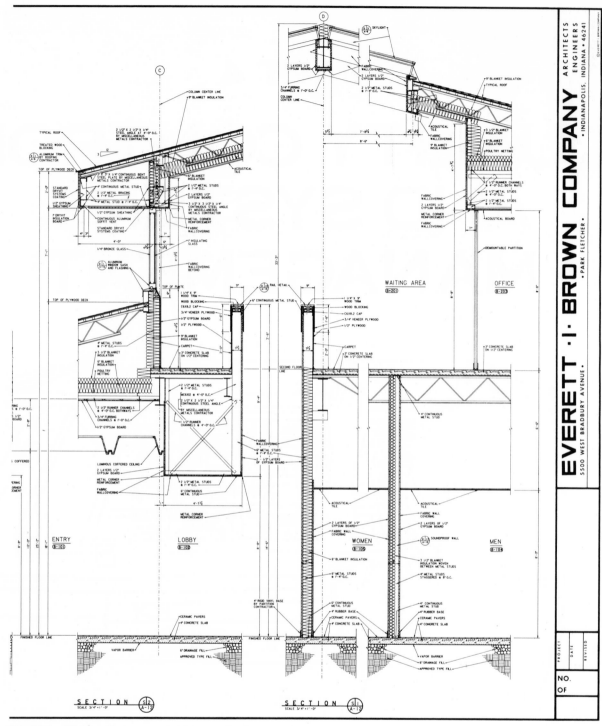

Fig. 14-2. Wall sections as created by a service bureau CADD system.

(Courtesy of Everett I. Brown Co.)

PART 4

QUALITY CONTROL AND SYSTEMS MANAGEMENT

Illustration by Lanée La 'shagway.

15

DISASTER PREVENTION

HOW PROJECTS FAIL: LESSON ONE

Some projects are doomed from day 1. Here's how a typical disaster project starts and how it worsens nonstop, even when everyone knows it's a disaster.

It usually starts with client irrationalities. The client's budget, expectations, or both may be unrealistic. Or the program is flawed. Or the client's predesign "sketches" are loaded with errors. The architect's or engineer's first task is to untangle these problems to clear the way for a rational design process. But a glitch develops.

The client becomes embarrassed and resentful, especially if some absurdities are included in the program. As they're revealed, the client (or client representative) becomes emotional. "No, that's not irrational. That's the way we want it. That's the way it's going to be if you intend to work for us."

The job has gone sour from the start, but the architects or engineers decide to tough it out. They rationalize: "Things are slowing down; we need this to keep our people busy." "The client's too important to turn down." "We can straighten them out as we develop the design."

Every step that follows is a replay of the first one. Program flaws are built in to the early design development drawings "just to see how it works out." "We'll pick those things up later," says the project director. "Later" is postponed through the design development and presentation drawings, and all the way into the working drawings.

Drafters get delays and no solutions every time they raise questions. Supervisors say, "We'll bring that up at the meeting. I'll get back to you on it." "Have the consultants look at it and let us know what they think." "Leave off those notes; we'll be able to pick all that up in the specs."

Contradictions and design errors have been passed from management to designers to drafters. The building has spaces that are too small for their functions, awkward circulation, illegal corridors and exits, conflicts between structure and mechanical work, and, by this time, a totally unknown construction cost. No one knows who will finally resolve the job's contradictions, but everyone believes that he or she is no longer personally responsible.

Meanwhile, there's an endless series of changes. Some original problems have become too obvious to ignore. The client has no memory of early demands and countermands them. "Move those doors. I can't imagine why you would allow traffic through those spaces anyway."

The project manager agrees to the minor changes without argument, having been instructed to keep the client happy. Some items are changed, changed again, and then reversed a step or two all in the course of a single meeting. Among all the scribbled notes, there's no single record as to what the final change decisions are, or why they were made, or who made them.

Each change creates a chain reaction of other changes. No one predicts the ultimate consequences of "minor" changes. Moving a few doors leads to relocating some equipment, which reduces some utility space, which involves the electrical and mechanical consultants. Construction costs are increased all out of proportion to just "moving a few doors."

Then comes the bidding. A disaster. The client is outraged at the cost overrun and orders redesign at the architect's or engineer's expense. The client also challenges previous billing from the design firm. "All these changes! And the overtime! Who on earth authorized these things?"

After bidding come the substitutions. A sales rep knows a much less expensive roofing system. "It's economical, and it's held up beautifully for years over at the shopping center." Unfortunately this roofing is not exactly the same as the shopping center's. "This metal curtain wall will go up at half the cost and in half the time." Unfortunately it's not compatible with the original structural system and will later buckle and split as the building frame expands and contracts.

The story goes on, of course, all the way to the courtroom. How can it happen? How is it possible for mature, educated, professional people to let a project disintegrate like that?

One answer goes like this: There's a psychological process called "evasion." Children do it frequently when reality conflicts with their wishes. They'll yell and scream denials of what is obvious and true. Adults use evasion as a defense. To admit to a certain fact would be painful, threatening, or embarrassing, so they blank it out of existence. Evasion takes on a life of its own in group activities. Everyone jointly avoids thinking out the situation or mentioning it. The unwritten rule is that the first person to name the really sticky problems gets to assume blame for having helped let them get out of hand.

I've named a problem—evasion. How about a solution? Is there a solution to something so deeply rooted in human behavior? There is one overall approach to management that nips the evasion problem in the bud. The approach explicitly favors clarity. It holds clarity and the exchange of tangible, usable information as the whole point and value of communication. Clarification destroys evasion. Listed below are a few tools and procedures to clarify job situations. They help prevent evasion and all the self-deceptions, secrecy, and many shades of lying that go with it.

1. A detailed "scope-of-services" or "scope-of-work" list. Most contracts include a general scope-of-work clause, but the list I'm describing is a separate special checklist to review point by point with the client before signing the contract. It lists all design firm services, with space for the architect or engineer and the client to record estimates of time and cost for each service. Later, if and when the client asks for added services or revisions, out comes the scope list for update and renegotiation. This clarifies one of the most common sources of client and architect-engineer misunderstanding.

2. Diagnosis, predesign, and a project program. This starts with a formal yes/no list of questions about job conditions and considerations. It zeros in on financial and legal analysis; project area; building configuration; logical structural, construction, and appurtenance systems; and spatial relationships. It resolves in advance what most designers spend weeks trying to figure out while doing schematics. Without formal predesign, major and minor decisions will be delayed through the working-drawing stage—even into construction, at higher and higher cost with every stage of delay.

3. Formal approvals of design decisions. This goes beyond the normal, overall design review and approval. It requires a room-by-room walk-through of the entire project with the client. Room by room, drawing by drawing, every point is either approved or not approved. If a drawing is not approved, possible changes are outlined. When a drawing is approved, the client initials the particular drawings.

   There's no better way to prevent misunderstandings or to block arbitrary client changes than a walk-through with formal approvals of every drawing and every room. This process is mandatory in several top design firms in the country. They consider it one of the best project time- and cost savers in existence.

4. Decision documentation, every step of the way. This is like the old-time "job diary," a day-by-day record of all design and production decisions on a job. It notes who makes a decision, when it is implemented, when it is revised, and who implements it. Above all, it notes the reason, the "why" of each decision. That's most important. Even if the reasons for a decision seem crystal clear and self-evident at the moment, they won't be clear later on. If you fail to keep notes, especially on decisions emanating from the

client, you will be held to account for any unexpected consequences.

I recently saw firsthand what happens when you fail to document decisions day by day. A major architectural and engineering firm faced huge dollar losses because the client was challenging the designers for some costly decisions. The client had no memory of participating in those decisions. Principals, managers, designers, and drafting supervisors were kept busy 15 hours a day for over a week and two weekends trying to reconstruct the decision sequence. And they didn't succeed.

BUILDING FAILURES AND QUALITY CONTROL

Roofs cost about 2 percent of construction budgets and bring in 50 percent of the lawsuits, according to the National Roofing Contractors Association. Roofs are famous as the number 1 point of failure, but we have a new contender among the trouble spots: walls. Exterior walls, once way down the list, are now the number 2 source of building failures. Exterior wall failures sparked 33 percent of all claims in 1980, according to the Illinois liability underwriters Shand, Morahan and Co.

Meanwhile, there's the annual epidemic: flooded paving and playgrounds, chronic masonry efflorescence, cracked slabs, window pop outs. The crazy thing is that the primary causes of these failures are well known and have been for years. The information is there; it just isn't getting to designers, drafters, and drawings.

Here's how the most quality-conscious offices handle the problem, using seven key elements:

1. One person to coordinate the detail system, document checking, and overall quality control. This is a part-time job in some firms, full-time in larger offices. It requires a committee in the largest organizations. The point is this: If no one is actually personally responsible for total office quality control, it will be left to the project managers on a project-by-project, day-to-day basis. That means everyone will reinvent the same data over and over and that officewide quality control will be uncoordinated and wholly unpredictable.

2. Standard details. Besides being a major time- and cost saver, a standard, reference, or master detail system allows an office to create and retain the very best of construction practice. That helps prevent less-qualified people from reinventing and, most importantly, misinventing those critical roofing, flashing, waterproofing, fenestration, and wall details. An active, constantly updated master detail system is a priceless office asset. But it doesn't happen if no one person is responsible for running it.

141

3. Independent checking of the documents. This means that a qualified person other than the project manager or job captain reviews drawings completely. This checking is separate from the normal phased check print and submittal process. It's solely a hunt for possible construction problems. Many firms schedule independent checks at roughly 10 percent and 80 to 90 percent of completion. At those times an independent checker can find problem spots that may be invisible to designers, drafting staff, and supervisors. Caution: the independent checker has to be construction-wise, knowledgeable regarding all aspects of detailing, specifications, and consultant coordination, and not have bones to pick with those doing the drawings. Otherwise the checker will nitpick the project to death.

4. Overlay checking of consultants' drawings. Whether you are using pin register overlay drafting or not, the best coordination tool consists of a light table and translucent prints of all the engineering trades. Checking the combined images of drawings of structure, plumbing, HVAC, etc., in one stack is far more reliable than leafing through separate check prints. Problems caught at this stage are infinitely cheaper to fix than they are during construction.

5. Combined specifications and drawings. As offices turn to offset printing, they're also adding reduced-size typed specifications—segmented to match the working-drawing divisions. (Smaller firms do the same by photocopying both sides of 11-inch by 17-inch working-drawing sheets. That size option is available on many plain-paper office copiers and is sufficient for smaller-building working drawings.) When drawings and specifications are combined as one package, they have to be cross-checked and tightly coordinated. That's not always the case with traditional documents.

6. The "postoccupancy" survey. Good for detail updating and design review, this walk-through survey is done well after a project is completed. Problems, corrections, and preventives are noted. Survey results are incorporated into the office manual and the detail system. As with all other aspects of quality control, someone has to be in charge and have the power to enforce postoccupancy surveys as office policy. (There's always strong resistance to looking for problems and lots of buck-passing when they're found.)

7. The final common element I've found in successful quality control programs is strong and clearly stated management support. Running a quality control program doesn't make friends, and it takes a strong, diplomatic manager to work through the resistance.

SYSTEMATIC CHECKING

The essence of a quality control or quality assurance program is the creation of office standards and more frequent and rigorous checking of drawings. Checking procedures are easy to change and can provide multiple benefits. Here are some good old and new suggestions:

Inspection and checking are always worst on the last batch of work to go through. Boredom or last-minute rushing leads to cursory checking at the very time inspection should be most thorough. This is when it pays to have someone who's not deeply involved in the job do the checking. ("Third-party" checking is done at 80 or 90 percent completion in some firms to avoid last-minute surprises just before printing for bids.)

You'll get very useful job information by elaborating a bit on your checking color coding. Most offices use a red marker for corrections and new information. Then a yellow marker is used to cross off the red mark items as they're taken care of on the drawings. You can point out drafting room problems by adding to the color system. For example, use red markers to show errors, overdrawing, and underdrawing. Use green to call out new data to be added and "excusable" omissions. Later, a review of job check prints will show up drafters or teams with excessively high errors and overdrawing or underdrawing rates. It will also show up checkers who are either too casual or excessively nitpicking.

Use a special color to separate client revisions from in-office additions, revisions, and refinements. Revisions in drawings required by changes in the client's program should be clearly marked to avoid legal and payment tangles later. If such changes are numerous, the check prints should be retained in the archives along with the original final drawings.

Drafting staff will hunt endlessly for check print sheets if sheets aren't kept in order and if there's no checkout control system. Use a binder clip to keep check print sheets together rather than stapling, unstapling, and restapling a set. And provide "file out" cards—8½-inch by 11-inch or larger sheets of poster board which are inserted whenever a check print sheet is removed. Drafting staff should note on the file out cards the date the check print sheet is removed and who has it. You can avoid an awful lot of searching and general aggravation by also enforcing a file out card or list control system for job tracings.

16

CONTROL AND COORDINATION WITH SYSTEMS MANAGEMENT

HOW YOU'LL DO A DOZEN TASKS IN A SINGLE STROKE

What's the next big change in architectural and engineering practice? Computerization? Of course. But the *real* change isn't the equipment; it's the new operating procedures that go with it—especially the thought processes they require. The procedures and processes give extraordinary productive power to the individuals who use them.

Some architects in particular say no to computerization. They write me to say it can't work in their highly diversified architectural practice or that it's impractical or too expensive for the very small firms.

Meanwhile, others—including the smallest of architectural firms—call or write about their extensive and growing computer operations. It's a technology gap that's widening in a hurry. While some traditional firms insist the change is years away, other firms of similar type and size take the "years away" technology for granted as an indispensible part of day-to-day routine.

What does it mean to say the real change is in the procedures and thought processes? Here's an example in several steps of what to expect—an elaboration on the Systems Management Documentation process described in Chapter 10.

1. Imagine yourself designing a project in a whole new way. Step by step you make decisions—either tentative or definite—on various aspects of the project. When you are doing site work, the decision might be a simple choice on how to handle site drainage or a choice of a brick privacy wall instead of a chain link fence. As you do so, you mark a code symbol which names the reason–tentative or definite—for the choice, such as "appearance—client request 5/10/83."

2. The decision you made doesn't stop there—the process of inputting that decision creates dozens of other decisions, definite and tentative, all at the same time. The choice of a brick privacy wall, for example, when entered in the computer, instantly identifies and prepares for printout the identifying note that will go on the site plan, and on any other appropriate drawings. If the wall is attached to the building and has to appear on exterior elevations, then a note is automatically assigned to those drawings, as is the brick pattern and the detail reference symbols that will appear on the exterior elevations.

3. Other data ready for printout from that one decision about the wall include:

Relevant standard details and outline keynotes to go with the details

The pertinent outline-specification section

Sources of technical literature and catalogs pertaining to such walls

Names and addresses of local suppliers of bricks and related materials

Applicable local building code or zoning regulations controlling heights, setbacks, or other aspects of design and construction

Cost data upon input of wall materials and dimensions

Other job consultants or building trades whose work may be affected by the decision

4. A working-drawing checklist is created and stored during the decision-making process which will be printed out later as a check on all drawing and coordination data that's to be included in final documents. The content and distribution of base and overlay sheets are similarly recorded for later management use. Also recorded is an outline PERT or CPM chart that includes drawing-by-drawing time and cost estimates for producing the drawings and specifications.

5. Some decisions create the need for others, or for early action. If you are deciding on a very special product—such as a special brick color and texture—a memo is automatically written to check availability from the manufacturer. When a decision has to be left tentative, or postponed entirely, it is earmarked to arise later as a question at an appropriate date. If a decision depends on someone else's consideration, a form letter or memo is automatically created for printout to ask for the decision, or for data affecting the decision.

6. When a decision is later considered for a change, all the repercussions and all the other previous decisions that will be affected will be immediately reported. When a change is made, it is noted and dated just like the original decision, and all related data, from notes to standard details to construction cost, are reset automatically.

Printouts of the whole process are periodically submitted to the client and consultants to check any inconsistencies or misunderstandings that might creep through.

This process means a radical change in the power of the individual. A drawing or note made in design development won't be remade in production drawing—you'll use the same data throughout the job. You won't have to write separate memos of inquiry or instruction. They're created during decision making.

You won't have one person noting the wall on a site plan, another noting it on elevations and making up a brick pattern, another creating or looking up details from scratch or making up detail notes, another looking up technical or product literature, another writing specifications, and so on. All that work so often duplicated by different individuals at different times—often in minor or major contradiction to one another—is done at once, inputted by one person and thus available for reuse at many levels of the system. All the data that have to be changed if the client later decides on a chain link fence instead of a wall—usually changed in a rush by different people on different documents—are mainly changed at one time, by one person.

At the end of the job, you'll have complete project history documentation, every step recorded from client contact to after final payment approval to the contractor. All this is created automatically, part and parcel of the yes-no-maybe decisions made throughout the job: what went into the documents; who made what decisions and when and why; an automatic costing of both design and construction changes, plus automatic costing of extra work for the architect or engineer when created by the client. And if the client gets confused as to what the final decision was about that fence or wall, the history of decisions and changes can be tracked through all the documents. And the design firm will *not* have to accept undeserved responsibility for client misunderstandings or memory lapses as to who decided what during the project.

Actually, there will be far more to all this than what I've just listed. There will be interface on the front end with the scope-of-services list, predesign, and automated building planning. There will be interactive options where, after predesign, the computer will make most of the obvious decisions about materials, details, and specifications, and you'll confirm, deny, or edit them. And there's the graphic interaction, designing with electrons instead of pencil and paper, and computer-drawn working drawings.

How long before all this becomes reality? Not long. I created the first aspects of this whole system in 1982, and most of the remainder will be completed during 1983 and 1984.

What kind of hardware will it take? Not much. *Most* of what I've just described can be done with "personal" or business-level microcomputers already commonly used for word processing and accounting. What I've described is software—the machines that can power it are all widely available—but the architectural and engineering software in particular has been comparatively slow in coming.

17

THE KEY
TO QUALITY CONTROL

CONSTRUCTION DETAILS ARE THE KEY

Details are the focal point in the quality control programs of hundreds of U.S. design firms. There are four main reasons why:

1. "Most construction failures occur in the details—either in design, drawing, or construction," says architect-engineer and building-failures expert Ray DiPasquale. He makes the point that while engineering students, for example, learn moment diagrams and how to use shear and load tables, the main danger points are in the joints, the detailed connections. A knowledgeable Cambridge, Massachusetts, engineer quoted in *Engineering News Record* concurs: "A large percentage of the time, collapses seem to be related to details rather than large structural principles." He adds: "It's the odd detail that you don't do everyday that gets you into trouble." This doesn't mean you stop inventing new details. It just means taking care—extra care through a systematic master detail system and a coordinated system of working-drawing project management and quality control.

2. Details, when coded by the CSI numbering system, can become an integral link that helps tie together every part of design and construction, including broadscope draw-

ings, notation, schedules, specifications, and project decision-making documentation.

3. A systematic detail-management program involves every aspect of systems drafting and graphics: paste-up, overlay, ink, typing, keynotes, computerization, etc. And a detail system provides the perfect lead-in for introducing all such systems in a simple, methodical way that both builds upon itself and pays for itself as you go along.

4. Besides quality control, architects and engineers achieve great time- and cost savings from master detail systems. An innovative residential architect, for example, gets 30 to 50 percent of detail reuse from his files. He originally didn't expect any reusability because of the diversity of his practice and the novelty of his designs. He learned, as have many others, that fundamental construction remains constant despite highly divergent design details. Thus, floor drains, roof drains, insulated drywall partitions, roof scuttles, metal ladders, cavity walls, etc., are found in only a very limited variety and are mainly repetitive from building to building. Larger-office users of standard or master detail files often report that 80 percent or more of the details on a complex detail sheet may come directly from the master files. That's an immense savings over time. And, it's a savings achieved by improving, not diminishing, design and construction quality.

THE ESSENTIALS OF A DETAIL SYSTEM

I've spent years researching detail systems and creating the beginnings of what I hope will become a national master detail system, and there's one thing that emerges loud and clear: most working-drawing details are repeats. They're copies, to one degree or another, of previous details. Those that aren't copied are researched and redrawn from scratch time and again in office after office. Whether your drafters are copying old details or reinventing them, it's a huge waste of time and money.

I'll elaborate on all the specifics of creating and running a master detail system shortly. First, here's a synopsis of the main components of such a system and how they come together.

First is a reference detail library collected from past jobs, other architectural and engineering firms, and published manuals and catalogs. Details are sorted and filed by category in sort of a custom, in-house version of *Graphic Standards*. The details cover all kinds of construction, especially unusual, one-of-a-kind conditions. They're for help when the "one-of-a-kind" circumstance comes up again. Importantly, when you go to an integrated master detail system, you get multiple use of every single detail that will ever be drawn in the future. Those that aren't suited to a master or standard system for direct reuse are still at least useful in the reference file.

Second, in addition to reference details, comes a file of generic or prototype details. These may be like our Guidelines detail starter sheets that include only major elements that are most repetitive and leave off the minor items most likely to vary from job to job. Ed Powers of Gresham & Smith sends their new generic details to construction-related trade associations for review and comment. That's an outstanding effort to record the very best construction information available and make it an integral part of design documentation.

The third quality control tool: master details. These are more specific details—mostly originated for ongoing projects by elaborating on starter sheets or generic details. They're left partly undone, copied, and the copies later re-viewed for inclusion in the master detail file. If they don't quite cut it for the master file, they go to the reference file.

The fourth tool, a jobsite feedback form, keeps the detail system fully up to date. If there's a troublesome detail discovered on a project, project representatives are required to send a brief memo back to whoever is in charge of the quality control–master detail system. The memo names the detail-type file number (which is printed with title and scale with every detail used in drawings). The system manager looks up the detail, and, while identifying the problem, looks at the detail drawing's detail history form. That's a space on the detail format sheet that identifies where else and when a detail has been used. The combination of jobsite feedback form and detail history form uncovers problems, allows quick corrective action on other projects that may be using the detail, and constantly screens and upgrades the entire master file.

There's more on all these points in the chapters that follow. They contain the best data on the subject I can find these days and can help your design practice immensely.

The Appendixes of this book include two comprehensive detail file number indexes for filing and retrieving construction details. The numbers are coordinated with the CSI Masterformat, and you'll find them to be an extremely useful time-saver as you establish or update your own master detail file.

18

HOW TO USE A MASTER DETAIL SYSTEM

THE DIFFERENCE BETWEEN REFERENCE DETAILS AND MASTER DETAILS

Reference details are those you might review for guidance in designing for some unfamiliar construction situation. They might be from previous jobs, from other offices, from product manufacturers, or from guide books such as *Graphic Standards* or *Time-Saver Standards*.

Most offices don't keep a formal reference detail file. They may keep useful details from previous work, but the details are usually not filed or indexed in any systematic way. Systems drafting–oriented offices use reference details increasingly these days. They're an excellent supplement for any master or standard detail file.

Master details represent standard construction. They show the repetitive conditions and are designed to be used directly—with or without revisions—in working drawings. Master details commonly include site work conditions, door and window details, standard wall construction, and connections of manufactured items such as roof drains and metal ladders. Even small offices sometimes compile and use as many as 5000 details showing common construction situations.

Besides being a major time- and cost saver, master details allow you to create and store the very best of construction practice. Offices that maintain well-managed detail files report a noticeable decline in problems at the jobsites. And there are definitely fewer problems with the buildings after construction.

Keep in mind the difference between reference details and master details. Reference details are a source of data in creating new original data. Master details are directly reusable drawings of common repeat items of construction.

CREATING THE DETAILS

People don't have to be taken off other work to create an office master detail file. You can create a master file by doing all future details in such a format that if they can be used in the master file, it'll be easy to enter them in it.

Details created in this fashion are drawn as usual but in a special format on 8½-inch by 11-inch sheets. All details are later reviewed for potential reusability in future projects. Even if a detail isn't acceptable as a master detail, most likely it can still be useful as part of the reference detail file. Either way you gain additional usage and value from most details long after they've been drawn for their particular project.

If you follow the *Guidelines* format, you'll be creating your future details within an approximate 5¾ inch high by 6 inch wide "window" on 8½ × 11 format sheets. The format sheets will show how to size and locate the detail title, how to align lettering, etc. When everybody uses the same format, details assembled on carrying sheets will look consistent.

The general rule in creating new details for future projects will be to split the drawing process into two steps:

1. Bring the detail to a point of *near* completion. "Near completion" means to leave off the material indications, dimensions, and notes that might vary in different circumstances. Some opening sizes or fabrication sizes might be variable, for example, so they're not specified at this step. The idea is to avoid putting so much information on the detail that it loses its potential reusability. (See Figures 20-1 and 20-2.)

2. After creating the detail up to a point of potential reuse, make two copies. Proceed to finish up your work for the job at hand on one copy. Add those variable material indications, dimensions, and notes that are necessary to complete the detail. The other copy is to review later for its potential use in the reference or master detail files. The original is filed separately. If a detail is selected for the master file, the separate original provides a backup—insurance against possible loss of the file copy. (See Figure 20-6.)

The copies will be sepia line diazo polyester reproducibles. If the original is clean, clear, and crisp, the reproducibles should be the same. The reproducible copies should have little or no background haze and no loss of line quality.

Each detail accepted either for the reference detail file or for the master detail file will receive a file number. This number will be integrated with the office's master specification numbering system. File numbers, and a master file number index, provide slots or "address numbers" for all filed details. It makes for the most convenient retrieval and cross-referencing of details after they've been filed away.

FINDING AND RETRIEVING DETAILS FROM THE MASTER FILE

As just described, each filed detail receives a permanent address or file number. The number identifies the detail's construction division relative to specifications. The detail may be numbered according to dominant material or by function. The specification number system determines exactly what numbers to use to identify the major divisions and their subcategories, or the broadscope and narrowscope. (See the Detail File Index in Appendix A1.)

The division numbering is reflected in the actual physical detail file. The file drawers contain hanging "Pendaflex"-type folders. Each folder is marked with a specification division name and broadscope number. Within each folder are smaller folders that contain the subcategories of details within a particular broad division. For example there might be a large folder identified as "Site Work—Division 2." Within that might be individual folders for "Curbs," "Parking Bumpers," etc. Each of those folders has a CSI-related number. Then the details within a folder each have their single identifying number. Thus a detail might be numbered 02528-4. Following the CSI format, that number would mean "concrete curbs within Division 2 site work." And the number following the hyphen would mean that this detail is the fourth concrete curb detail in the file.

It would be inconvenient to poke through a lengthy index of detail names and numbers to find a detail that you hope might be on file, so you need a convenient cross-reference system. One that shows what details are available and tells how to find them in the file.

The cross-referencing, or "lookup," system is in the form of a three-ring binder—a detail catalog. The detail catalog is a filing system ancillary to the file number structure of the file folders. Instead of being shown in the catalog in the way they're filed, details are shown in the sequence that people would be most likely to look for them. If you want to check out some window details, instead of looking through separate divisions of steel windows, aluminum windows, and wood windows—all separate specification sections—you'd look in the general category of the catalog labeled "Exterior Walls." Then you'd search for the alphabetical subsection under "W" for "Windows." Within that you could look for further subdivisions of window types and materials. (See the Detail Catalog Division Index in Appendix A2.)

When you locate some details in the catalog to try out, note their file numbers and have the masters retrieved and copied for you. Then make up either a detail book or a large-sheet-size paste-up of the masters. Add special data unique to your project to complete the details, finish the paste-up if you're doing paste-up, and proceed with check prints.

RETRIEVING DETAILS FROM THE REFERENCE DETAIL FILE

If you're working out some new details dealing with data not on file in the master system, check the reference detail file for help. Reference details will be filed according to the same index number system as the master details. To avoid confusion of the reference details with master details, some firms start reference file numbers with a distinctive "R."

If the reference detail library is very large, details will be kept in file folders similar to the master file just described. If it's practical, however, it's desirable to keep reference details in a binder—the office's own version of *Graphic Standards*.

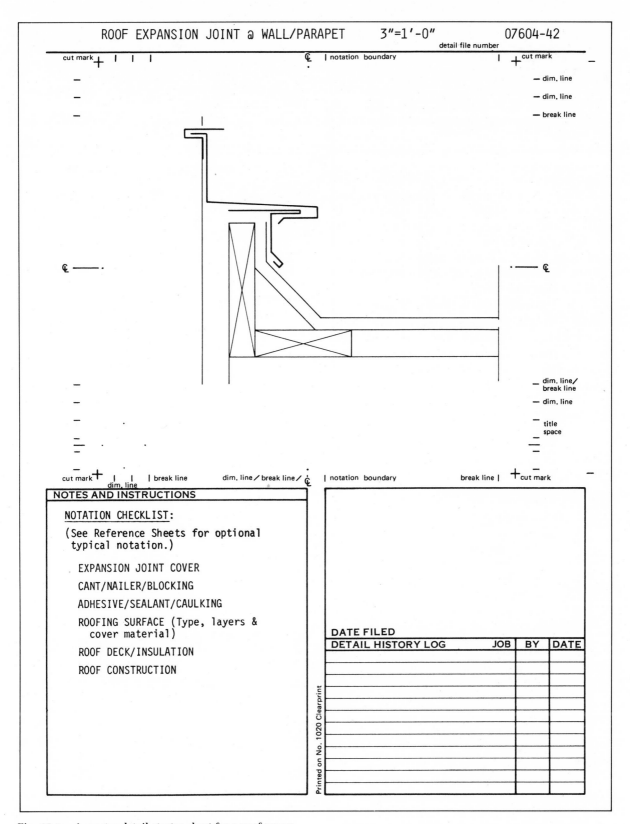

ROOF EXPANSION JOINT @ WALL/PARAPET   3"=1'-0"   07604-42

detail file number

cut mark ✛ | | | ₵ | notation boundary | ✛ cut mark   —

—   — dim. line

—   — dim. line

—   — break line

₵ — .   . — ₵

—   — dim. line/
                                                                                  break line

—   — dim. line

—   ⁻ title
                                                                                  space

—

cut mark ✛ | | | break line   dim. line / break line / ₵ | notation boundary   break line | ✛ cut mark   —
          dim.
          line

NOTES AND INSTRUCTIONS

NOTATION CHECKLIST:

(See Reference Sheets for optional
 typical notation.)

 EXPANSION JOINT COVER

 CANT/NAILER/BLOCKING

 ADHESIVE/SEALANT/CAULKING

 ROOFING SURFACE (Type, layers &
   cover material)

 ROOF DECK/INSULATION

 ROOF CONSTRUCTION

DATE FILED

DETAIL HISTORY LOG	JOB	BY	DATE

Printed on No. 1020 Clearprint

Fig. 18-1. A master detail starter sheet for a roof expansion joint. Only generic data are shown.

(Guidelines.)

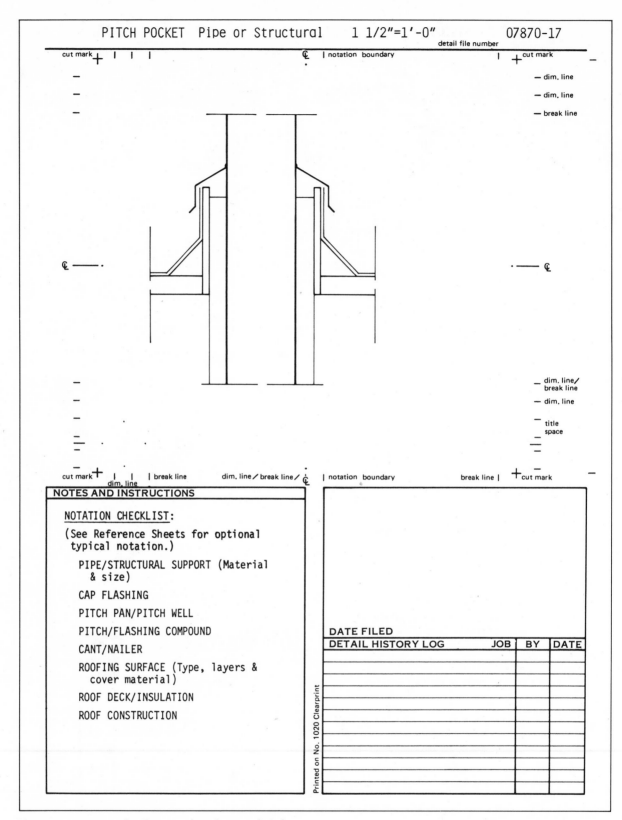

Fig. 18-2. A master detail starter sheet for a roof pitch pocket. Like the roof expansion joint, it only shows general generic data. Specifics are to be added by the user.

(Guidelines.)

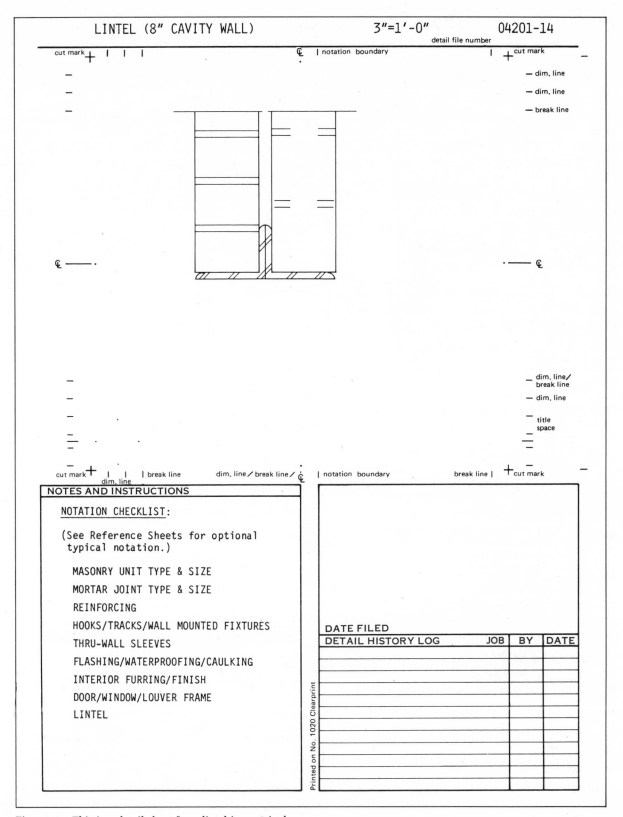

LINTEL (8" CAVITY WALL)   3"=1'-0"   04201-14

detail file number

cut mark + | | | ℄ | notation boundary | + cut mark —

— dim. line

— dim. line

— break line

℄ —   · — ℄

— dim. line/
break line

— dim. line

— title
space

cut mark + | | | break line   dim. line / break line / ℄ | notation boundary   break line | + cut mark —
dim. line

NOTES AND INSTRUCTIONS

NOTATION CHECKLIST:

(See Reference Sheets for optional
typical notation.)

MASONRY UNIT TYPE & SIZE

MORTAR JOINT TYPE & SIZE

REINFORCING

HOOKS/TRACKS/WALL MOUNTED FIXTURES

THRU-WALL SLEEVES

FLASHING/WATERPROOFING/CAULKING

INTERIOR FURRING/FINISH

DOOR/WINDOW/LOUVER FRAME

LINTEL

DATE FILED

DETAIL HISTORY LOG	JOB	BY	DATE

Printed on No. 1020 Clearprint

Fig. 18-3. This is a detail sheet for a lintel in an 8-inch
cavity wall. This detail starter can be combined with door
bucks, windows, or any other appropriate data to form
any number of additional details.

(Guidelines.)

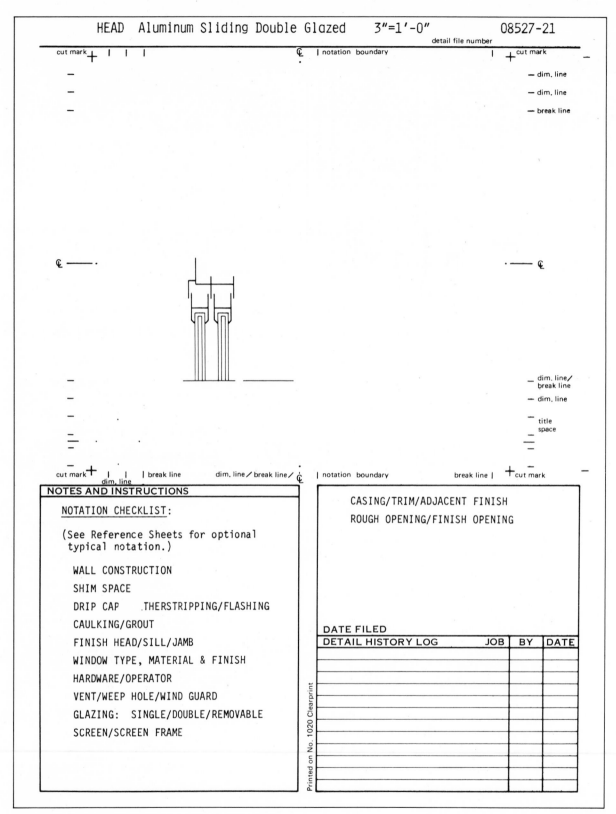

HEAD   Aluminum Sliding Double Glazed   3"=1'-0"   08527-21

detail file number

cut mark   notation boundary   cut mark

— dim. line

— dim. line

— break line

Ȼ   .— Ȼ

— dim. line/
break line

— dim. line

title
space

cut mark | break line   dim. line ∕ break line ∕ Ȼ | notation boundary   break line | cut mark

dim. line

NOTES AND INSTRUCTIONS

NOTATION CHECKLIST:

(See Reference Sheets for optional
typical notation.)

WALL CONSTRUCTION

SHIM SPACE

DRIP CAP   .THERSTRIPPING/FLASHING

CAULKING/GROUT

FINISH HEAD/SILL/JAMB

WINDOW TYPE, MATERIAL & FINISH

HARDWARE/OPERATOR

VENT/WEEP HOLE/WIND GUARD

GLAZING: SINGLE/DOUBLE/REMOVABLE

SCREEN/SCREEN FRAME

CASING/TRIM/ADJACENT FINISH
ROUGH OPENING/FINISH OPENING

DATE FILED

DETAIL HISTORY LOG	JOB	BY	DATE

Printed on No. 1020 Clearprint

Fig. 18-4. This is an aluminum sliding window starter
sheet, strictly generic, which can be combined with wood,
masonry, or other wall sections.

(Guidelines.)

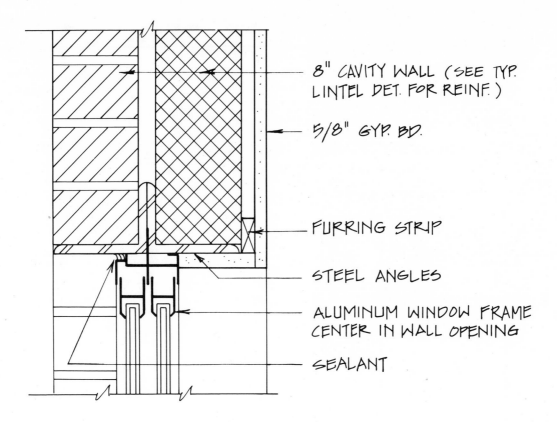

8" CAVITY WALL (SEE TYP. LINTEL DET. FOR REINF.)

5/8" GYP. BD.

FURRING STRIP

STEEL ANGLES

ALUMINUM WINDOW FRAME CENTER IN WALL OPENING

SEALANT

WINDOW HEAD (JAMB SIM.)

SCALE: 3" = 1'-0"

08527-21

Fig. 18-5. This shows the preceding starter details combined as a whole new detail for a standard detail system. The office has added texture or materials indications, notation, leader arrows, and title according to its office standards and drafting style.

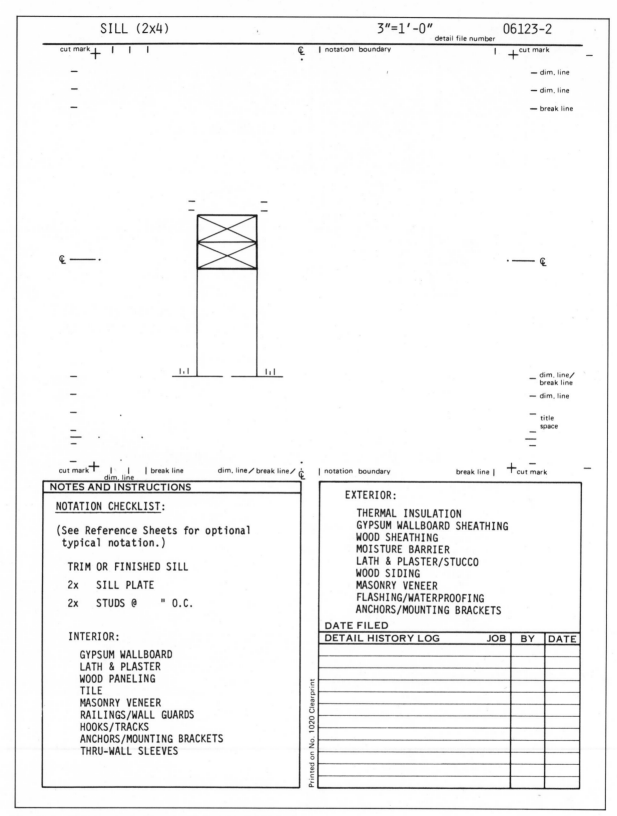

Fig. 18-6.   This is a typical wood wall framing detail that can be combined with other detail starter sheet data.

(Guidelines.)

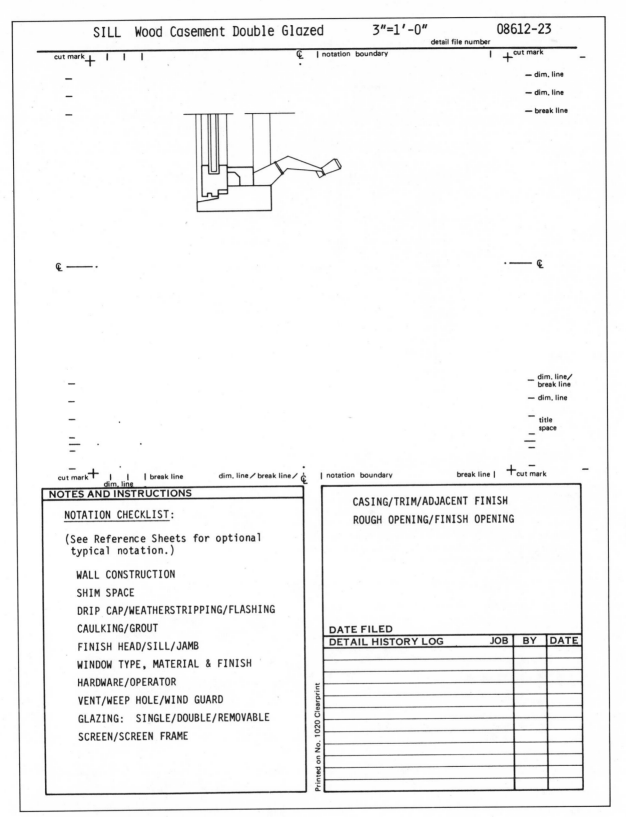

SILL Wood Casement Double Glazed 3"=1'-0" 08612-23

detail file number

cut mark | notation boundary | cut mark

— dim. line
— dim. line
— break line

ℂ ℂ

— dim. line/ break line
— dim. line
— title space

cut mark | break line | dim. line / break line / ℂ | notation boundary | break line | cut mark
dim. line

NOTES AND INSTRUCTIONS

NOTATION CHECKLIST:

(See Reference Sheets for optional typical notation.)

WALL CONSTRUCTION

SHIM SPACE

DRIP CAP/WEATHERSTRIPPING/FLASHING

CAULKING/GROUT

FINISH HEAD/SILL/JAMB

WINDOW TYPE, MATERIAL & FINISH

HARDWARE/OPERATOR

VENT/WEEP HOLE/WIND GUARD

GLAZING: SINGLE/DOUBLE/REMOVABLE

SCREEN/SCREEN FRAME

CASING/TRIM/ADJACENT FINISH

ROUGH OPENING/FINISH OPENING

DATE FILED

DETAIL HISTORY LOG	JOB	BY	DATE

Printed on No. 1020 Clearprint

Fig. 18-7. This is a wood casement window, as a starter sheet detail that can be combined with divergent wall sections.

(Guidelines.)

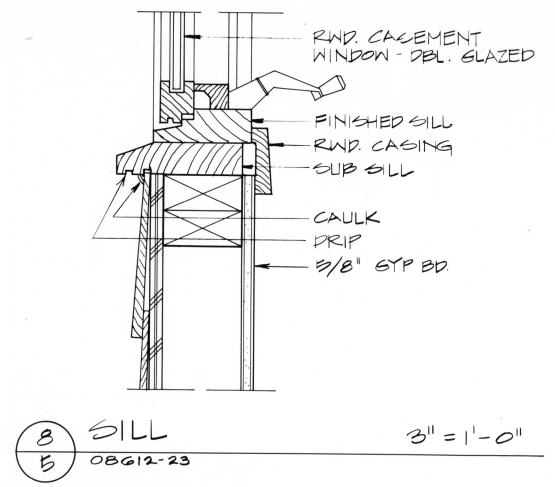

RWD. CASEMENT
WINDOW - DBL. GLAZED

FINISHED SILL
RWD. CASING
SUB SILL

CAULK
DRIP
5/8" GYP BD.

SILL   3" = 1'-0"
08612-23

Fig. 18-8.   The window detail and wall framing detail are combined either by paste-up or by overlay drafting. Materials indications, connective pieces, wall finishes, etc., are added in as required for the job at hand. Now the detail is complete and can be numbered for the master or standard detail file or for the reference detail file.

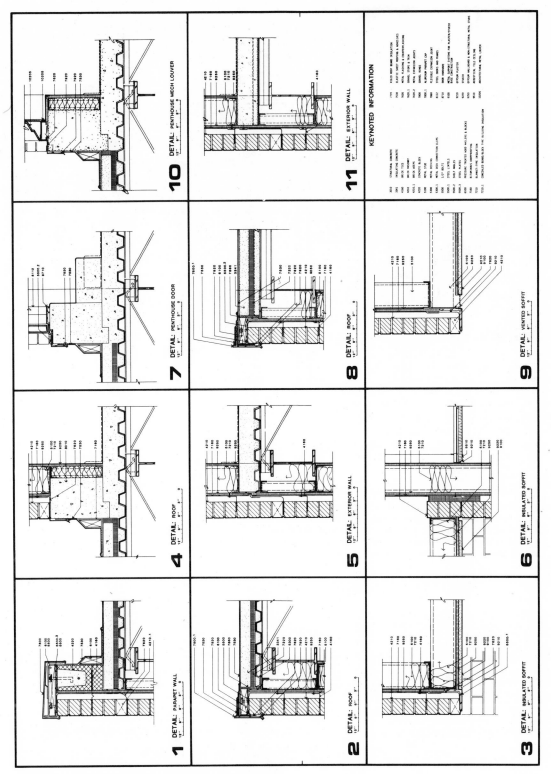

Fig. 18-9.   Another version of adaptation of generic details in a standard detail system. The components, like components of construction in general, are common and can be combined and recombined on drawings just as they are in the building construction process. (*Illustration turned on page.*)

(Courtesy of Ed Powers, Gresham, Smith and Partners.)

161

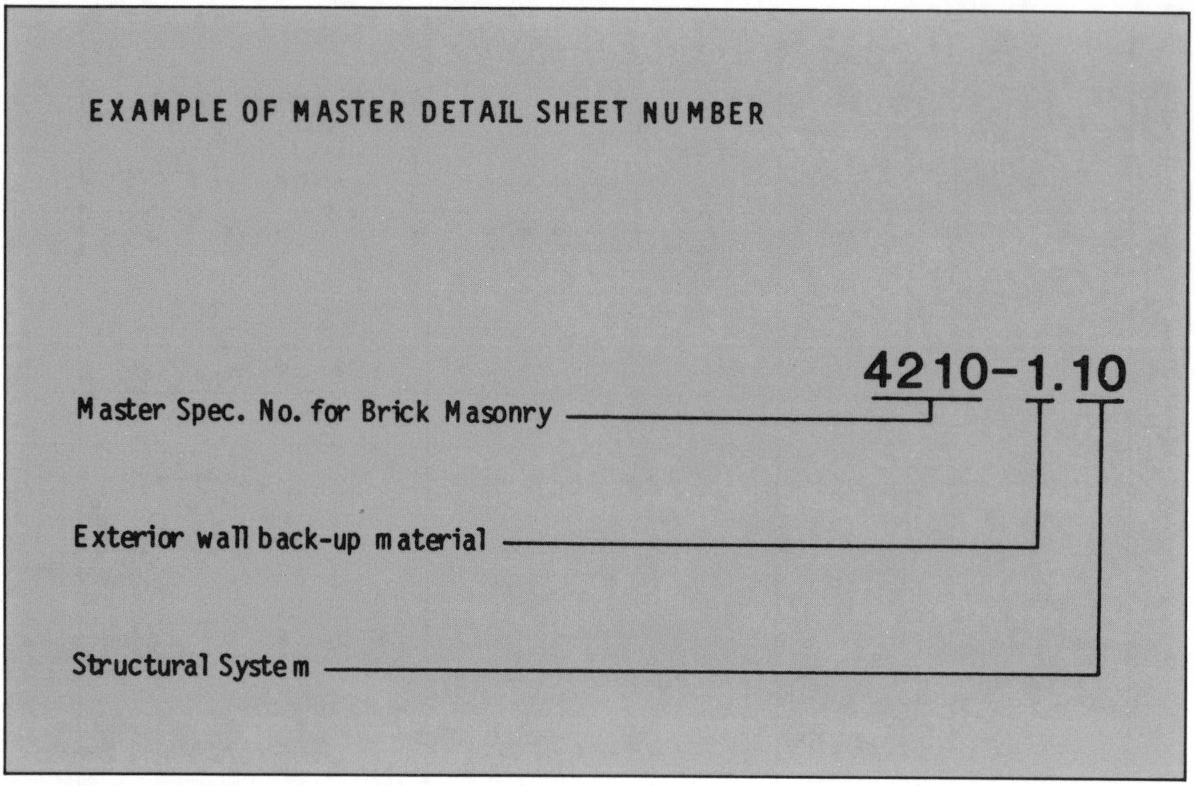

EXAMPLE OF MASTER DETAIL SHEET NUMBER

4210-1.10

Master Spec. No. for Brick Masonry ————————————————

Exterior wall back-up material ————————————————

Structural System ————————————————————————

Fig. 18-10. The notation for these details is standardized and referenced by keynotes similar to the system described elsewhere in this text.

(Courtesy of Ed Powers, Gresham, Smith and Partners.)

19

CHECKLISTS FOR CREATING CONSTRUCTION DETAILS

Here are the most widely tested and accepted procedures and standards for creating original construction details.

RULES FOR DESIGN AND SKETCHING

_____ **1**. Details and detail types should first be listed as part of a miniature mock-up working-drawing set. Draw cartoon miniatures of plans and elevations, then indicate bubble keys at joints, junctures, and typical sections where details will be needed.

_____ **2**. Make miniature cartoons of needed details on one-fourth-size mock-up sheets to show their approximate size, positioning, and coordination with one another. Name the details.

_____ **3**. Search your standard or master detail file for possibly usable details for the project at hand. Make office-copier prints and assemble them on a paper carrier sheet as a rough check print.

_____ **4**. Search your reference detail library for details that will aid in design of exceptional and nonstandard details. Assemble office-copier prints onto carriers for review.

_____ **5**. Search the office technical library for published details, such as product manufacturer and trade association literature and drawings.

_____ **6**. Proceed with rough design sketches on 8½-inch by 11-inch sheets as necessary to devel-

op new original details. Copy the wholly new details and assemble the check print copies on carrier sheets.

_____ **7**. Use red pencil or pen to show revisions, additions, and deletions on the full-size detail check print sheets.

This essentially completes the research and planning groundwork.

RULES FOR USING SYSTEMS GRAPHICS TECHNIQUES FOR COMPLETING DETAILS

_____ When creating new details, such as wall sections, watch for opportunities to use common background data as base sheets while using overlays to show the unique detail situations.

_____ Watch for opportunities to photo-reduce oversized, freehand-sketched details instead of redrafting them from scratch.

_____ Watch for opportunities to copy common detail elements in quantity for creation of base detail information as paste-ups.

_____ Use a standardized 8½-inch by 11-inch detail format sheet identical to your standard detail file format sheets for _all_ new detail final drawings. Remember, new details that aren't suited for direct reuse in

163

your standard file are still reusable as sources of ideas and technical information in a separate reference detail file.

____ Set details within a consistent cut-out, paste-up "window" (6 inches wide by 5¾ inches high is a preferred window size for paste-up modules and detail format sheets).

____ Plan on using typewritten notation and keynoting wherever practical.

RULES ON DRAFTING STANDARDS FOR CONSISTENCY AND CLARITY

Besides exhibiting general sharpness and high contrast, contemporary design and working drawings have to be designed for special reproduction processes. Elements may be photo-reduced, for example, so elements in the original drawing have to be extra large to remain readable. Or drawing components may be copied through several generations of reproduction, so they have to be extra crisp, clear, and black to begin with, or they'll fade away. Traditional grays, pochés, and "light and dark" lines just don't cut it anymore.

____ All hand lettering must be large, a minimum of ⅛ inch high, with 3/32- to ⅛-inch spacing between lines of lettering. Major titles should have a minimum of ¼ inch high lettering; minor titles should be a minimum of 3/16 inch high. Typewritten notation should be in uppercase, pica size. Smaller letters tend to break up or fade out.

____ Line work must be consistently black and vary only by width, not by "darkness" or "lightness." Light lines or gray lines disappear during printing.

____ Use dots or lines, not grays, to indicate materials. Grays don't reproduce properly with the new graphic repro systems.

____ All symbols such as arrowheads, feet and inch marks, and circles must be large and unmistakably clear. Small symbols tend to clog up during reproduction.

____ Line work or crosshatching must be spaced at least 1/16 inch apart; otherwise, lines will run together in printing.

____ No poché or other drawing is allowed on the back of the drawing sheet.

____ Don't let line work touch letters or numerals. Don't let fractional numbers touch their division line. That will avoid blobbing.

____ Ink line work on polyester drafting media is the preferred media combination. Pens for use on polyester must be jewel- or carbon-tungsten-tipped. Use ink erasing, fluid–imbibed erasers such as the Pelikan PT 20.

RULES ON GRAPHIC LAYOUT FOR CONSISTENCY AND CLARITY

____ Draw from the general to the particular and from the substructure or structure outward to the finish materials.

____ Draw in layers, so that a detail drawing is substantially a completed composition at every step along the way.

____ Draw in sequence from the primary to the secondary—detail component first, size dimensions second, notation third. Add textures, crosshatching, and profiling last.

____ In general, keep the exterior face of construction facing to the left, interior to the right.

____ In general, notation should be in a column down the right-hand portion of the detail window. Some notes can and should be located elsewhere if necessary to avoid crowding or to enhance clarity.

RULES ON NOTATION

____ The primary purpose of notation is to identify the pieces of the detail by generic material name, construction function, or both. Leave all extensive material descriptions, brand names, tolerances, and construction standards to specifications. On a small job that combines specification information with notes, follow the above rule but add general or "assembly" notes, as a short-form specification on the drawing sheets.

____ In general, a note states the size of the material or part first, tells the name of the material or part second, and names the position or spacing third. If using keynoting, the keynote legend will identify the names of materials or parts, and the size, and sometimes the spacing will be tagged onto the keynote reference bubble in the field of the drawing.

____ Follow office guides on nomenclature and abbreviations. The office should follow recommended guides of the professional societies and print their versions of such guides as part of working drawing general information.

RULES ON DIMENSIONING

____ Avoid fractions where possible. When using fractions, never use those, such as ⅓ inch, that can't be measured with normal scales. The smallest practical fraction in most dimensioning is ¼ inch.

____ Don't draw dimension lines to the face of material. Unless you have no other option, draw only to lines extended from the material or detail part.

____ Add terms such as "hold," "min.," "max.," "N.T.S.," "equal," "varies" to state your intent clearly on the accuracy or absoluteness of your dimensions.

____ Avoid double dimensions that show a size on one side of the detail and repeat the sizing on the other side.

RULES ON LEADER LINES FROM NOTES TO DRAWING ELEMENTS

____ Preferred line style is straight-line, starting horizontally from the note and breaking at an angle to the designated material or detail part.

____ Curved lines are acceptable if preferred as office style, but they should be done with french curves or large circle templates, not freehand.

____ Don't use leader lines from a note to more than one detail or section.

RULES ON LINE WEIGHTS

____ Break lines, dimension lines, and cross-hatching are narrowest. General background construction is medium width. The primary detail object should be "profiled"

with the widest line. Equivalent pen tips would be 000 or 00 for narrow lines; 0 to 1 for medium; 1, 2, or 3 for profile lines. Exact choice of line widths is variable depending on the scale of the detail. The wider lines look excessive in small-scale details, and the lesser line widths look weak with the larger-scale details.

____ Remember, lines are to be differentiated by width, not lightness and darkness. All lines should be completely opaque for best reproducibility.

RECOMMENDED SCALES FOR CONSTRUCTION DETAILS

The commonly used scales for architectural working-drawing details are:

$\frac{3}{4}'' = 1'0''$

$1\frac{1}{2}'' = 1'0''$

$3'' = 1'0''$

The ¾-inch scale is used for the most elementary details such as simple residential or light-frame construction footings and fireplace sections, plain standard cabinet sections, and simple landscaping details. A 1-inch scale is used for similarly simple details, but most firms go to a 1½-inch scale when the ¾-inch one is not adequate.

The 1½-inch scale is recommended for most construction situations. It's widely used for all types of building components such as walls, floors, ceilings, roofs, and related items. It is the dominant detail-drawing scale used in architectural working drawings.

The 3-inch scale is also widely used for numerous common conditions such as door and window sections, and for numerous small items such as brackets, tracks, curtain wall connectors, handrails, and thresholds. Sometimes this is a matter of judgment. In borderline cases, it's considered best to go to the larger size to assure clarity.

One-half-size and full-size details are useful only for extra small connections or detailed design studies or shop drawings. They are time-consuming and usually don't show more data or show it more clearly than is possible with 3-inch scale details.

Similarly, details in ¾-inch and 1-inch scales should be discouraged because if the drawing becomes at all complex, detail of these scales won't convey the information clearly.

The ½-inch, ¾-inch, and 1-inch scales are acceptable for wall sections. Take care that the sections aren't so large that they just duplicate the information on the larger close-up wall construction details.

COMBINING YOUR MASTER DETAILS WITH A QUALITY CONTROL SYSTEM

There are two main tools for introducing quality control to your master details:

1. The jobsite feedback form
2. The master detail format sheet with a detail history log.

A sample jobsite feedback form is shown in Figure 19-4. Use this form, or your version of it, for quality control communications from the site. Whenever a site representative discovers a problem with a construction detail in action, it must be reported in this fashion. This will allow the person in charge of your detail system to modify and upgrade the detail in question. Without this kind of feedback, you'll find the same details causing the same problems in project after project.

The master detail format sheet with a detail history log is a multipurpose tool. Its role in quality control is to show what jobs use what details, where they're used, when, and by whom. If a problem comes up, other projects that may be under way using the same detail can be corrected. It will also show which details are used frequently and which are not used. This shows the manager of the system which details to augment and which to move out of the active master file into the reference file.

Other tools of quality control include checklists of details that should be included on all types of drawings and an independent quality control check of each set of working drawings. Such special checking processes usually occur at about the 10 and 80 percent stages of project completions.

THE MASTER DETAIL FORMAT SHEET

Figure 19-1 shows a master detail format. This is the kind of format you'll want to have offset-printed onto fade-out blue-grid tracing paper or onto gridded polyester. The intent is that your staff will henceforth create all new construction details in a consistent format and size that can be conveniently coordinated in large-sheet paste-ups.

Some firms don't print their format on 8½-inch by 11-inch sheets as shown. Instead they provide "template" underlay sheets to guide staff members as they draw details on blank sheets. When they later print the detail for filing and cataloging, they combine the detail drawing with the clear-film template—like base and overlay—and print the combined images together on final masters.

Whether you preprint the format or use tracer sheets, and whether details are drawn on paper or polyester, it's imperative that all details henceforth be drawn in the 8½ × 11 size, within a consistent size module, and on a no-print blue-grid medium. The grid helps tremendously in normal drafting and is mandatory when the designer or drafter is drafting freehand or using tapes or other appliqués.

Everyone will ultimately find the 8½-inch by 11-inch size most convenient for drafting—far more convenient than sprawling over large-size drafting sheets. But it's a change, and some will inevitably find reasons to resist an unfamiliar drafting method. They'll argue, for example, that certain details that relate to one another, such as portions of wall sections, should be drawn on a larger sheet rather than be divided on separate small sheets. The argument is correct regarding the design and rough sketching out of the details and regarding final composition of the details on a carrier sheet. But actual hard line drafting of each final detail is measurably easier and faster on smaller sheets. And, of course, using the small-sheet format is the only way to generate usable reference and master details for your filing system.

The illustrated detail format sheet can be modified any way you see fit. You might prefer to fit the details and note spaces sideways on the sheet. You may want to use a different module than the 6 inch wide by 5¾ inch high module recommended here. Whatever specific design you come up with, I mainly recommend that you consider including all the kinds of information shown: cut lines, suggested dimension and note boundaries, title limitations, detail history log, reference note space, etc.

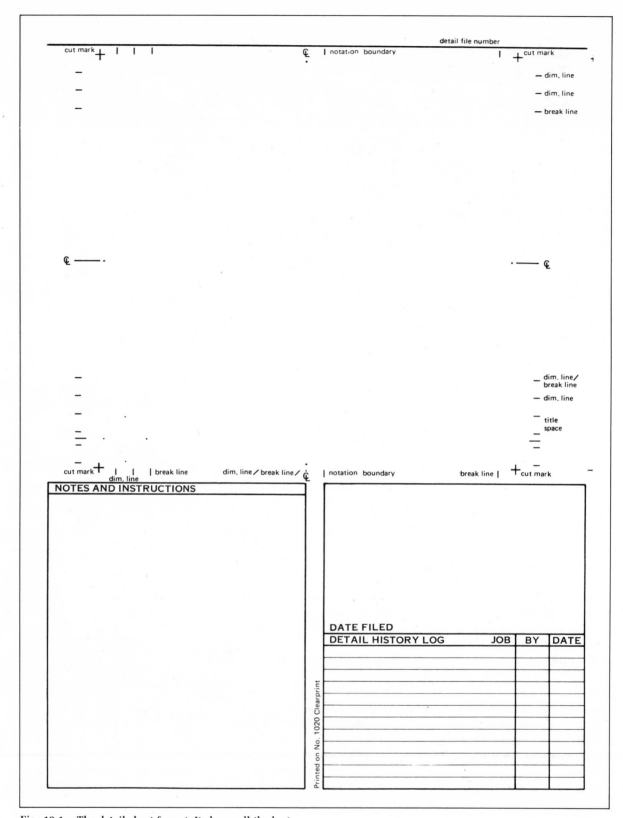

Fig. 19-1.   The detail sheet format. It shows all the best positioning of detail information.

(Guidelines.)

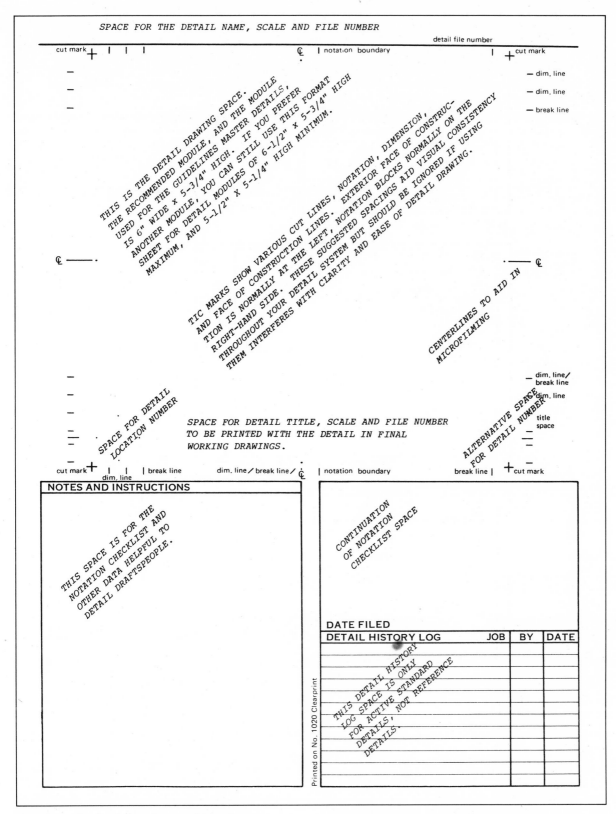

Fig. 19-2. The detail sheet format with explanatory
notes.

(Guidelines.)

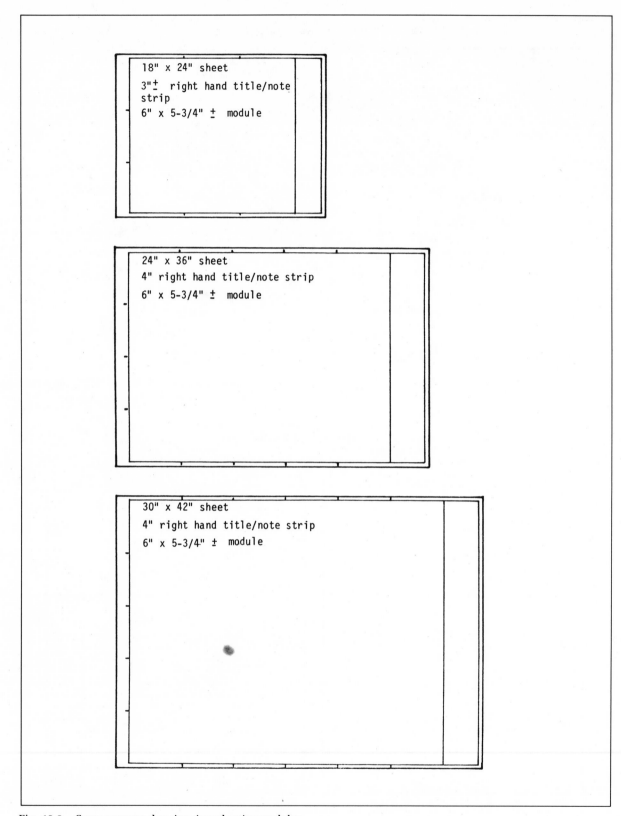

Fig. 19-3. Some common drawing sizes showing modular divisions based on the recommended master detail sheet cutout window module. The window is 6 inches wide by 5¾ inches high. All sheets use a 1½-inch left-hand binder margin and a ½-inch border around the remainder with a title block and keynote strip on the right.

Jobsite Feedback Form

[This form requires an 8½″ × 11″ sheet, your letterhead, and any necessary items from your standard submittal form.]

Jobsite feedback submittal to: _____

Copy to: _____

Date: _____

Job name: _____

Observer's name: _____

If problem involves specifications, note division and section number:

If problem involves large scope working drawings, note sheet number:

If problem involves construction detail(s), note:

Job sheet number _____

Detail location number on the drawing _____

Detail file number (if any) _____

Description of design or construction problem:
(Use back of sheet if you need more space,
and add photograph if possible.)

Possible or likely cause of the problem:
(Use back of sheet if you need more space.)

Fig. 19-4. A jobsite feedback form which helps keep the standard detail system constantly updated with data from the jobs as they're being built and from data from postoccupancy surveys.

20

STANDARD DETAILS AND IN-HOUSE REPROGRAPHICS

WHY USE TRANSLUCENT PASTE-UP?

The technique described here saves more offices more time and money than any other single aspect of systems drafting. I'll describe it in the context of creating detail drawings, but it's 100 percent applicable to any kind of drawing.

Why use any kind of paste-up? Paste-up or composite drafting allows you to take advantage of the key feature of systems drafting—reusability. Reusable elements in working drawings range from door swings to ceiling grids, from repetitive rooms to repetitive fenestration. The point of creating a master detail system, of course, is to exploit the commonalities and repetition that inevitably occur in construction detailing. You store common elements so people don't have to draw and redraw them time and again. You store what's common, then later make copies, add new information, assemble the copies on carrier sheets, and finally make reproducible prints of the paste-ups.

Early efforts at paste-up required the use of photography. A paste-up would be photographed onto a same-size negative. The negative image would then be transferred to a draftable, translucent, plastic-base photo medium. This was expensive and cumbersome, but still sometimes proved to be a major time- and cost saver. I'll comment on improved, contemporary photo systems later.

Other variations on paste-up techniques arose as new equipment and media appeared on the scene. In the early 1960s, 3M sold a thermal copier stickyback product that let you copy directly from existing drawings and transfer stickyback copies onto new tracings. Copy quality wasn't adequate, and the stickybacks left very dark shadows on prints, so the product was abandoned. Most offices have tried to use their office copiers in a similar fashion and have encountered problems of distorted size, uneven line quality, ghosting, and difficulties in mounting, changing, or removing the copier stickybacks.

All along there's been a great need for products that would make good copies, could be assembled conveniently as paste-ups, and would not leave ghost shadows in printing. We now have such products in three forms: (1) transparent films usable with certain special office copiers; (2) photo-washoff material that can be used in room light with vacuum frame printers; (3) sepia line diazo polyester film. Most design firms find that the sepia line polyester films provide the greatest reliability and versatility at the lowest cost.

THE MEDIUM YOU NEED FOR SUCCESSFUL PASTE-UPS

I've just stated that sepia line diazo polyester film is your best bet as a reproduction medium for paste-up. Let's examine just what that product is.

First, the base material, polyester, is what's popularly known as "Mylar." Mylar is Du Pont's trade name; polyester is the generic name. Polyester is favored over tracing paper in systems drafting work because it is strong and stable.

Polyester is available as a drafting medium, as a photographic material with silver emulsion, and as a diazo print medium. Diazo printing is the common ammonia-developed print process most architects and engineers use to produce check prints and job prints. Don't confuse it with photo printing or with a nonammonia print chemistry such as Bruning's PD 80.

Diazo emulsion on polyester may be black, sepia, red, green, blue, and other colors. The color best-suited to reproduction is sepia. The sepia line emulsion may appear reddish brown, yellow-brown, or dark brown. Printed sepia line work may look translucent in some instances, but it definitely is not. As emphasized early in the text, the sepia color is engineered to be especially opaque—sort of "super dark"—when reproduced in a diazo print machine. The sepia color is particularly resistant to the ultraviolet light used in diazo print machines.

A sepia line reproducible print can be so intensified that it will actually make a darker, clearer line print than the original drawing, especially if the original was done in pencil. This is an impressive capability—to be able to make a "second original" that's actually better than the real original. It's impressive and extremely valuable.

One value is that you can do an original drawing in pencil on tracing paper, make a sepia reproducible copy, and have the copy make better prints than the original. Similarly, you can type notes on tracing paper, copy the tracing on a diazo print machine onto sepia polyester, and end up with a darker and more printable set of notes than you started with.

The intensifying quality of diazo sepia films can enhance original drawings, but it won't make up for deficient line work. Final print quality is always ultimately dependent on the darkness, crispness, and clarity of the originals. That's why ink drafting on polyester is favored by most systems-oriented offices.

The most commonly used brands of sepia line diazo polyester are James River Graphics, Precision Coatings, Ozalid, Arkwright, Keuffel & Esser, and Teledyne. For catalogs of the varieties of diazo films available, contact the people listed in our *Reprographic and Computer Resources Guide* cited at the end of the book.

The preferred diazo film for making masters and paste-up reproducibles is 3-mil thick, draftable matte surface on the top side and erasable sepia emulsion on the back side. You may also want to use non-matte surface material, called "slicks," for special purposes. Slicks are used as printed base sheets in overlay drafting, for example, and for reproducing paste-up elements such as titles that won't require any further drafting.

HOW TO MAKE REPRODUCIBLES OF PASTE-UPS

Since the sepia line reproducibles are so intensely opaque to the diazo machine light, you can expose them in printing for a longer time than normal, without burning out the image. That means you can make a paste-up of sepia line reproducible elements, print it at a slower speed, burn out any traces of shadows, ghosts, or cellophane tape marks, and end up with a ghost-free sepia line reproducible print of the paste-up.

There are two basic conditions to meet: (1) Your original has to be high-quality so that the second-original masters are especially intense in line quality. (2) When making sepia print elements for paste-up, you must burn out as much background haze as possible.

The reproducible, if on draftable and erasable polyester, can be treated as an original tracing. Erase it, add new elements—whatever you want.

If you've burned out as much background haze as possible, you won't have to be concerned either about "white spots" from erasing or about cutting out portions of the reproducible polyester. There may be slight traces of "spotting," but they will disappear later when you do extended exposures to make final-stage reproductions.

Many people doing overlay drafting have worked with sepia line, clear, non-draftable surface polyesters as base sheets. They have seen such sepia images cloud up over time. So the natural question is: How long do the sepia line paste-up elements last? What happens over time to the sepia reproducible you make of the paste-up?

The answer has several parts. First, there are differences in products, and the better ones don't fog up nearly as much as the inexpensive, throwaway material often used as base sheet slicks in overlay drafting. Second, keep in mind that the paste-up elements are used as an intermediate step, not as a final, permanent work sheet. Third, if you want to help ensure long life for a sepia polyester film copy, you can rerun the developed print through the print light and developer twice, to burn off any residual emulsion that might be on the sheet. It's the ongoing developing of residual emulsion in the air that appears as fogging of the print surface.

I've seen quality diazo films remain clear for years, even after extended exposure to fluorescent light. If stored in light-tight cabinets and kept from heat or exposure to ammonia fumes, your sepia prints should last as long as you need. The final protection for keeping sepia copies over time is to periodically check them in the file drawers. If there's any sign of deterioration, reproduce them again. For 100 percent archival quality, of course, go to photo-washoff media at the close of a project.

ALTERNATIVE MEDIA FOR PASTE-UPS AND REPRODUCIBLES

One alternative to the diazo polyester film is Du Pont's Crovex contact duplicating film. Although a "washoff" product, it can be handled in subdued room light and can be exposed on a vacuum frame. Contact any local Du Pont sales representative or repro shop that uses Du Pont products for samples and literature.

Although we don't recommend office copiers for making paste-up elements, some firms are happy with them. To be usable, copiers must have extremely dark and consistent toning systems, be able to reproduce accurately at 1 to 1 size, and copy well onto clear copier films without jamming the machine. Copiers that have worked well for architectural and engineering firms include the older Royal Bond models III and IV, the newer Royfax 115, Xerox 2020 and 2080, A.B. Dick 990, and Kodak Ektaprint 100 and 150. Some machines require optical adjustments for greater size accuracy in copying. Some require that the fusing temperature be raised in order to copy onto clear polyester films.

In general I believe you're better off working with the diazo sepia films than with copier prints on clear films. If it happens you have a copier of exceptional quality and have worked out an acceptable copier film technique, then of course there's no reason not to continue with it.

I caution against the use of copier stickybacks as a means of reproducing and transferring details. Stickybacks and copier stickybacks have many uses, but most films and copiers do not meet the demands of large-image detail copying. For one thing, most toner and stickyback combinations are not erasable. Toners are often muddy. It's difficult to apply larger stickyback sheets quickly to a carrier and equally difficult to pull them off. The stickybacks readily jam up in copiers. For these and many other reasons, I'm convinced you're far better off just lightly taping 3-mil polyester films down on carrier sheets than struggling with stickyback products.

Some design firms have their own process cameras and developers and can produce photos on clear film in-house. Photo transparencies certainly provide the sharpest possible paste-up elements. If you can produce them conveniently in-house, there's no reason not to. If you also happen to have a projection camera and washoff processor, you may also prefer to do total photo reproduction rather than diazo paste-up and contact printing. It can be very convenient and economical to operate such a facility in-house, as opposed to depending on a distant repro shop.

Another repro option is provided by the Xerox 2080 repro machine. It has variable enlargement and reduction capability and will copy on polyester film. Its price, close to $80,000, is prohibitive for most firms, but if you have one, it will work well for your detail paste-up and reproduction system.

THE PASTE-UP PROCESS, STEP BY STEP

Here is a list of the steps for creating detail paste-ups and reproductions. I've focused on the diazo process, but most of the basic operations can be translated in terms of photo or copier reproduction.

1. Start with the original detail drawing, presumably done on an 8½-inch by 11-inch standard detail format sheet like that described elsewhere in the book.

2. As described earlier, you hold back completing original details. Just include information of potential general use, and exclude data that are highly particularized and most likely to be applied one time only.

3. When you reach the stop point in step 2 above, make two copies of the detail on draftable, erasable, sepia line film. Three mils is a good thickness. Make the reproducible prints by placing the drawing side flat against the emulsion side of the print medium. That helps ensure the closest contact during printing. The prints can be made on a regular diazo machine, if some slight size change doesn't matter, or on a flatbed contact printer vacuum frame. The print medium should be matte on the top side only, emulsion on the back.

4. File one copy of the original detail in a "holding file." Later it will be reviewed for inclusion in the office reference or master detail system. As indicated in Chapter 18, place the original in a separate file to serve as a backup in case the file copy is damaged or lost.

5. Proceed with finish-up work on the second, or master, copy as required to complete the detail for the job at hand. Any added line work should be in ink or plastic lead, to approximate the intensity of the sepia image already on the reproducible master.

6. Collect a group of details and any other drawing components to be composed on a single paste-up sheet. The composition of paste-up sheets should be preplanned in miniature working-drawing mock-up sets. If you're doing a whole sheet of details, use a large-grid module coordinated with your standard detail cutout size. A popular module is 6 inches wide by 5¾ inches high. That module fits well within the drawing sizes of 24 inches by 36 inches and 30 inches by 42 inches, when combined with a right-hand vertical 4-inch keynote strip and title block.

7. Cut the details to size to fit the paste-up sheet module. Assemble the details on a 3-mil transparent polyester carrying sheet. If the sheet size is 30 inches by 42 inches or larger, you may prefer to use a thicker, 4-mil sheet. Some offices prefer to do paste-ups on a draftable, matte-on-one-side carrier instead of a clear, non-matte one. Their thinking is that they may need to add some drawing directly on the carrier. Sometimes this is so, but most often it's preferable to consider all elements in a drawing as paste-up elements. If there may be changes while composing the sheet, use easily removed drafting tape at first to get elements into position. For final taping, use utility-grade cellophane tape or other strong clear tape. "Mending" tape tends to be hard to remove when making changes.

8. To make a check print of the paste-up, lay the original paste-up backside against the print paper and expose it on a vacuum frame or diazo printer. If running the check print through a diazo machine, protect the paste-up from damage with a thin cover sheet of clear polyester. Ghosting on a check print isn't particularly important, so you may choose to print at a higher speed than you would to make a ghost-free print.

9. To make a reproducible print on either paper or polyester, lay the original paste-up face down against the emulsion side of the print medium. If running this on a diazo machine, the paste-up portion will be automatically protected by the print sheet. You may have some difficulty with slippage if you try to run polyester sheets face to face. They tend to slip around during printing and create a fuzzy final print. For best print quality, of course, use a vacuum frame. Since you want your reproducibles to be ghost-free, use a print time long enough to burn out tape marks, shadows, edge marks, etc., but not so long as to burn out line work. If the original line work was not dark enough to begin with, it will start disappearing at this phase. You can allow minor ghosting and a few marks on the reproducible because this too can be burned out in later printing.

10. At this point you may face the decision of whether to do further changes directly on a **177**

reproducible of the paste-up or to hold off and do changes strictly on the paste-up itself. The rule is that if many changes are likely, you should hold off work on a reproducible until near the end of the job. Meanwhile, proceed with all revisions directly on the paste-up.

Those are your instructions for making in-house paste-up elements, paste-up sheets, and reproducible diazo second originals. Your final diazo job prints should be ghost-free and, in fact, print sharper than a traditionally hand-drafted tracing.

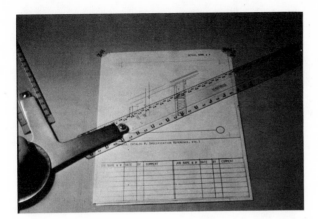

Fig. 20-1. This sequence shows the creation of details for paste-up and reproduction.

First, a detail is drawn to a point of near completion. That which is wholly variable and unique to the job at hand is left off the detail as it's brought up to a "copy stop point." (The copy stop point is explained in Figure 20-4.)

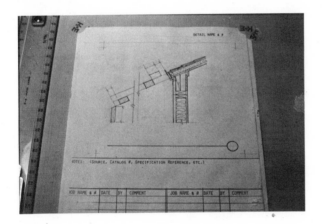

Fig. 20-2. The detail is created on 8½-inch by 11-inch tracing paper or drafting polyester. It has a cutout window 6 inches wide and 5¾ inches high. And it includes data forms which will be useful if the detail becomes part of an office standard detail file.

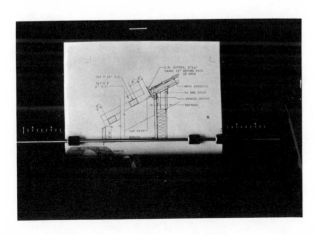

Fig. 20-3. Since the drawing is on 8½-inch by 11-inch format, it's easy to slip it in the typewriter to add in the notation. Typed notes are more legible, and they create a consistency in final graphic quality that's impossible to get if you have different drafters adding details in the same job or on the same sheet.

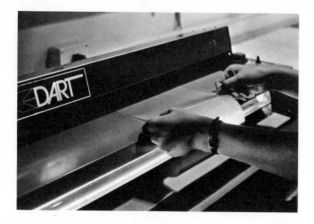

Fig. 20-4.   When a detail is judged to be done enough to consider for possible reuse in a standard detail library, that's its copy stop point. At this point two draftable reproducible prints are made. The original is set aside, and drafting that's unique to the job at hand is added to one of the copies. The other copy is considered for inclusion in the detail file. Here a sheet of sepia line polyester print media has been exposed to the detail and is being fed into the ammonia developer portion of the diazo print machine.

Fig. 20-5.   The rotary machine has a slight effect on size fidelity in printing, but that's not considered a problem in construction details. What may be seen as more of a problem is the slow speed required to make sepia diazo polyester copies. A vacuum frame may be preferred both for size accuracy and for speed of reproduction.

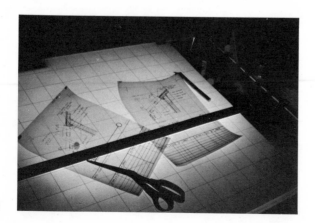

Fig. 20-6.   Here are two copies of the detail on the light table. They can be drafted on, erased, cut out, etc., as required.

Fig. 20-7. The detail copy which will be used for the current job is completed and checked. For good later reproduction, it's important that these copies have a minimum of background haze.

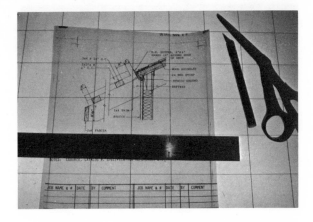

Fig. 20-8. The detail is cut to size for paste-up on a large working-drawing sheet. The detail can also be used in a detail book if the office chooses that route. Most offices prefer the paste-up method, however, partly because they like to show details that relate to one another in total context.

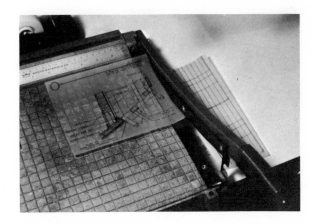

Fig. 20-9. When a detail is accepted for the office's standard or reference detail file, it will be photocopied on paper and displayed in the "office catalog." Drafters and designers sort through the catalog when planning their detailing to see what's on file that's reusable.

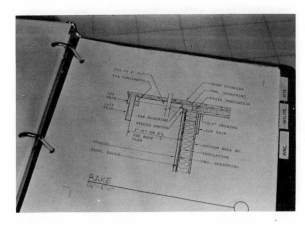

Fig. 20-10. When a detail is selected for paste-up, it's copied on the sepia line film and taped in position with transparent tape onto the polyester carrying sheet.

Fig. 20-11. Here's the paste-up of details. Details are taped only at the top corners. That way you can run a check print through a rotary diazo machine without the details getting bunched up and torn off the carrier. Note that pieces of details not wanted are just cut out. That's faster than erasing. If you erase and leave "white spots," they will be "burned out" in the final printing.

Fig. 20-12. It's generally best to hold off making a reproducible until the very end of a job and then do any final revisions on that sheet. Otherwise, keep all additions, deletions, and revisions on the original paste-up.

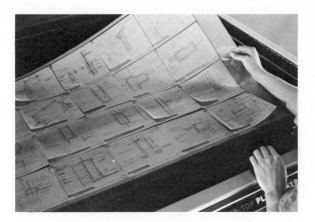

Fig. 20-13.   Now to reproduce the paste-up. It's a translucent paste-up, so you can print it in the same basic way as any translucent media drawing. Here, we'll reproduce it on the vacuum frame. First it's laid on the print sheet. The details are shown face up here for clarity, but in reality you'd set them face down onto the emulsion side of any reproducible print media. If you were making a blue line check print, the print sheet would be on the bottom and the paste-up atop it and face up as shown.

Fig. 20-14.   The lid is pulled down onto the bed of the vacuum frame. The operator has to watch for displacement of original and print sheet, for bent pieces of paste-up sheet, etc. Exposure times for various print media in combination with various types of originals have been recorded and are usually on a chart next to the vacuum frame.

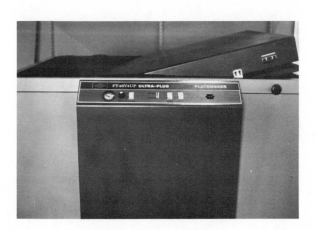

Fig. 20-15.   This is a "flip-top" vacuum frame. The glass top copy bed flips over so that the print sheet and original transparency material can be exposed to a powerful ultraviolet light source inside the box.

Other vacuum frames use light tubes. Some single-point light source units are top-lighted rather than having a flip-top design. The composition of your originals and print sheets will vary depending on whether the sandwich of sheets is lit from above or below.

Fig. 20-16. The exposed print medium has been developed in the ammonia developer portion of a standard diazo print machine. To further intensify the image and remove traces of background emulsion, rerun the sheet through the ammonia developer and then reexpose it to ultraviolet light. That process both finishes off development and then burns off remaining emulsion that might otherwise fog up the print as it developed in the air over a period of time.

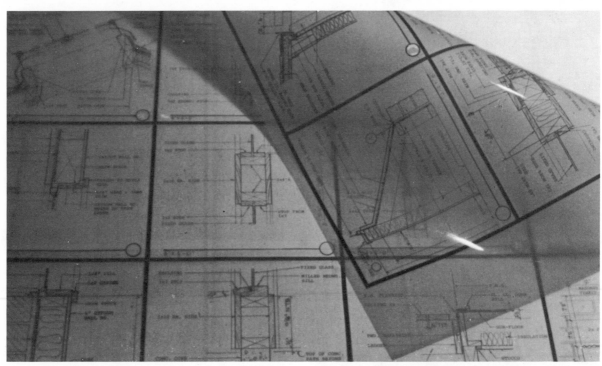

Fig. 20-17. This is the final reproducible. It has a draftable matte surface on the up side; erasable emulsion is on the back side. Since the original paste-up was of sepia line polyester, the line work was so opaque that the paste-up could be exposed to light long enough to burn off traces of tape marks, outlines of paste-up, and shadows or "ghosts." The reproducible is as clean as any original tracing would be, and since it too is on sepia line polyester film, it will print faster and with higher contrast than a traditional graphite-drawn paper tracing sheet would print when used for bid and job prints on diazo paper.

APPENDIXES

THE DETAIL FILE INDEX

EXPLANATION OF THE NUMBERING SYSTEM

This is the recommended detail file system. All reference details and all master details should be filed in a sequence comparable to this one. The file numbers are integrated with the Construction Specification Institute Masterformat. This facilitates direct coordination of detail drawings and specifications. It will also allow complete coordination of drawings and specifications with a master keynote system.

The first number group in each listing is the CSI division number for that particular material, assembly, or fabrication.

The basic CSI divisions included here are:

2 Site Work

3 Concrete

4 Masonry

5 Metals

6 Wood and Plastic

7 Thermal and Moisture Protection

8 Doors and Windows

9 Finishes

10 Specialties

11 Equipment

12 Furnishings

13 Special Construction

14 Conveying Systems

In a typical detail file, there will be a hanging file folder labeled Division 2—Site Work. Then, within that hanging folder there will be a separate file folder labeled 02400—Drainage. Within that subfolder there will be a collection of details filed by narrowscope CSI section numbers. In addition, there will often be clusters of details within a CSI narrowscope number, such as different kinds of foundation drains filed under 02411. When there are additional details under one CSI number, you have to add a hyphen, and then further numbers. Your third foundation drain detail on file, for example, would have as its final file number 02411-3. Your tenth foundation drain detail would be 02411-10.

Some other minor elaborations on the numbering system are described under "Notes" in the text of the index. See particularly the notes with Division 8—Doors and Windows.

IMPORTANT NOTE: MOST NUMBERS IN THIS INDEX ARE NOT INDIVIDUAL DETAIL FILE NUMBERS. INSTEAD THEY REPRESENT CATEGORIES OR SECTIONS WITHIN THE DETAIL FILE. THE FINAL INDIVIDUAL DETAIL NUMBERS WILL HAVE FURTHER SUBDIVISIONS OR SUFFIX NUMBERS DEPENDING ON WHAT SPECIAL CONDITION THEY SHOW, SIZES OR TYPES OF COMPONENTS, SCALE OF DRAWING, AND SEQUENCE OF SELECTION OF THE DETAIL FOR INCLUSION IN THE FILE.

DETAIL FILE INDEX

DIVISION 2   SITE WORK

DIVISION 3 CONCRETE

T-Type
Weep Hole
03303 PIER
Bell Pier
Socket Pier
03304 GRADE BEAM
Tie Beam
03305 FOOTING
Spread Footing
Stepped Footing
03306 FOUNDATION WALL (LIGHT-FRAME CON-STRUCTION)
03307 FOUNDATION WALL (HEAVY-FRAME CONSTRUCTION)
03308 SLAB ON GRADE (LIGHT-FRAME CON-STRUCTION)
Construction Joint
Control Joint
Depressed Slab
Isolation Joint
Raised Slab
03309 SLAB ON GRADE (HEAVY-FRAME CON-STRUCTION)
Construction Joint
Control Joint
Depressed Slab
Isolation Joint
Raised Slab
03310 CONCRETE WORK
03316 CONCRETE WALL
Areaway Wall
Construction Joint
Isolation Joint
Movement Joint (Expansion-Contraction)
Pipe Sleeve
Thru-Wall Opening
03317 CONCRETE COLUMN
03318 CONCRETE BEAM
Pipe Sleeve
03319 CONCRETE SPANDREL BEAM
03321 FLAT SOLID CONCRETE SLAB
Anchor
Control Joint
Curb
03323 FLAT PLATE SLAB
Depressed Slab
Pipe Sleeve
Raised Slab
Thru-Slab Opening
03325 METAL DECK CONCRETE SLAB
03327 ONE-WAY CONCRETE JOIST SLAB
03329 WAFFLE SLAB
03330 ARCHITECTURAL CONCRETE (ORNAMEN-TAL)
03350 SPECIAL CONCRETE FINISHES
03400 PRECAST CONCRETE
03411 PRECAST CONCRETE WALL PANEL
03412 PRECAST CONCRETE DECK
03413 PRECAST CONCRETE PLANK

Hollow-Core Slab
Solid Flat Slab
03414 STEMMED DECK
Double T
Single T
03415 PRECAST GIRDER
03416 PRECAST BEAM
03417 PRECAST JOIST
03420 PRECAST PRESTRESSED CONCRETE SECTION
03430 TILT-UP CONCRETE
03450 ARCHITECTURAL PRECAST CONCRETE
03500 CEMENTITIOUS DECKS
03510 GYPSUM CONCRETE DECK
03515 PRECAST GYPSUM PLANK
03530 CEMENTITIOUS WOOD FIBER SYSTEM
03540 COMPOSITE CONCRETE AND INSULATION DECK
03550 ASPHALT AND PERLITE CONCRETE DECK

DIVISION 4 MASONRY

NOTE

When you index and file masonry wall details, you may wish to subdivide them in the same way as the following breakdown of cavity wall details:

Wall at Grade

Thru-Wall Opening

Thru-Wall Pipe Sleeve

Header

Wall at Spandrel

Wall at Slab

Wall at Roof

Wall at Parapet

Wall at Existing Construction

This or a similar sequence could apply to brick or concrete walls of any type.

Many firms choose not to detail doors and windows with their wall details. Consequently we're referencing "sill," "jamb," and "head" details in Division 8—Doors and Windows.

04150 MASONRY ACCESSORIES
04160 JOINT REINFORCEMENT
04170 ANCHOR
04180 CONTROL JOINT
04200 UNIT MASONRY
04201 CAVITY WALL (BRICK AND CONCRETE BLOCK)
04204 CAVITY WALL AT SLAB
04205 CAVITY WALL AT SPANDREL
Shelf Angle
04206 CAVITY WALL AT ROOF
Shelf Angle

DIVISION 5 METALS

DIVISION 6
WOOD AND PLASTIC

06111 LIGHT WOOD FRAMING—EXTERIOR WALL
06112 PREASSEMBLED COMPONENTS
06113 SHEATHING
06114 STRUCTURAL PLYWOOD
06115 WALL AT FOUNDATION
06116 WALL AT FLOOR
06117 THRU-WALL OPENING
06118 WALL AT CEILING
    Soffit
06120 INTERIOR WALL FRAMING (Vertical sections)
06121 TYPICAL WALL FRAMING (Plan sections)
06122 JAMB
    Cased opening
    End wall
06123 SILL
06124 HEADER
06125 WOOD DECKING
06130 HEAVY TIMBER CONSTRUCTION
06132 MILL-FRAMED STRUCTURES
06133 POLE CONSTRUCTION
06150 WOOD-METAL SYSTEMS
06151 WOOD CHORD METAL JOISTS
06160 ROOF FRAMING
06170 PREFABRICATED STRUCTURAL WOOD
06180 GLU-LAM
06190 WOOD TRUSSES
06200 FINISH CARPENTRY
06220 MILLWORK
    Balusters
    Baseboards
    Casings
    Jamb Trim
    Wall Trim
06400 ARCHITECTURAL WOODWORK
06410 CABINETWORK
    Bathroom
    Commercial Laboratory
    Display
    Kitchen
    Shop
06411 COUNTER TOP
    Bathroom
    Kitchen
    Laboratory
06412 SHELVING
    Book Shelves
    Linen Shelves
    Utility Shelves
06413 WARDROBE STORAGE
    Flat Storage
    Luggage Storage
    Pole and Shelf
    Shoe Storage
06420 PANELING
06421 DECORATIVE WOOD PANELING
06422 PEG BOARD
06423 STORAGE WALL PANEL

06430 STAIR WORK
    Stair Section
06431 WOOD RAILING
06432 TREAD AND RISER
06433 STRINGER
06434 EXTERIOR STAIR FRAMING
    Posts and Footings

DIVISION 7 THERMAL AND MOISTURE PROTECTION

NOTE

Wall flashing at sills and heads of windows and exterior doors is normally included with the door and window details. See Division 8—Doors and Windows.

07100 WATERPROOFING
07150 DAMPPROOFING
07200 INSULATION
07210 BUILDING INSULATION
07220 ROOF AND DECK INSULATION
07230 PERIMETER AND UNDER-SLAB INSULATION
07250 FIREPROOFING
07300 SHINGLES AND ROOFING TILES
07310 SHINGLES
07320 ROOFING TILES
07321 CURVED ROOF TILES
07322 FLAT ROOF TILES
07400 PREFORMED ROOFING AND SIDING
07410 PREFORMED WALL AND ROOF PANELS
07420 COMPOSITE BUILDING PANELS
07461 WOOD SIDING
07465 PLYWOOD SIDING
07500 MEMBRANE ROOFING
07510 BUILT-UP BITUMINOUS ROOFING
07520 PREPARED ROLL ROOFING
07530 ELASTIC SHEET ROOFING
07570 TRAFFIC TOPPING
    Duck Board
    Roof Deck
    Sleepers
    Walkway
07600 FLASHING AND SHEET METAL
07601 DAMP COURSE
07602 WALL FLASHING
    Spandrel
    Thru-Wall Opening
07603 PARAPET FLASHING
    Cornice
07604 PARAPET FLASHING AND ROOF EXPANSION JOINT
07605 METAL FASCIA/FLASHING AT ROOF EDGE
07606 VENT FLASHING
07607 CURB FLASHING
07608 CHIMNEY FLASHING

07610 SHEET METAL ROOFING
07620 SHEET METAL FLASHING
      Sheet Metal Trim
07630 ROOFING SPECIALTIES
07631 GUTTER
07632 DOWNSPOUT
07633 SCUPPER
07660 GRAVEL STOP
07665 METAL COPING
07800 ROOF ACCESSORIES
07810 SKYLIGHT
      Curb
07822 ROOF DRAIN
07823 ROOF DECK DRAIN
07825 BREATHER VENT
07830 ROOF HATCH
      Hatch Ladder
07840 GRAVITY VENTILATOR
07850 PREFAB CURB
07851 CONCRETE CURB
07852 WOOD CURB
07860 PREFAB EXPANSION JOINT
07861 METAL EXPANSION JOINT
07870 PITCH POCKET
07871 GUY WIRE CONNECTION
07900 JOINT SEALANTS
07910 JOINT FILLER/GASKET
07920 SEALANT/CAULKING

DIVISION 8 DOORS AND WINDOWS

NOTE

Door and window details are among the most numerous in any reference or master detail file. The large number of possible details requires that we expand upon the CSI division numbers we're using as file numbers.

For example, assume you're establishing a file slot for some standard steel frame sections. Some are with concrete walls and some with masonry. Start with the steel frame number from the CSI, which is 08100. Then add a dash and a number to represent the wall construction according to CSI division. Concrete would be -3; Masonry -4; Metal Frame Wall -5; Wood Frame -6; Solid Plaster -92; Solid Gypsum Board -925; Tile Wall -93. Then add a hyphen and the final individual detail identification number.

A sequence of steel door frames in Masonry would look like this:

08111 (that's the CSI section for Standard Steel Door Frame)
  Plus -4 (the CSI division for Masonry)
    Plus -1, -2, etc. (for the details in sequence)
      Equals 08111-4-1, 08111-4-2, etc.

Remember, don't be concerned about what may seem to be overly long file numbers. The numbering is done only once, then copied automatically as details are copied and inserted in working drawings. The longer numbers are necessary in this part of the detail file and will *not* impose any time-consuming complexity into the system.

Use the same breakdown to file the window details. Identify window type and CSI number, add the wall type and the individual final detail number. You can cluster with the same file number separate drawings for head, jamb, and sill for the same window type in the same wall type. For example, a steel casement window would be 08512. If in a masonry wall, it becomes 08512-4. And if it's the ninth detail in your file sequence, it would be 08512-4-9.

08100 METAL DOORS AND FRAMES
08110 (STANDARD) STEEL DOOR
08111 (STANDARD) STEEL FRAME
08112 (CUSTOM) STEEL DOOR
08113 (CUSTOM) STEEL FRAME
08115 (PACKAGED) STEEL DOOR AND FRAME
08120 ALUMINUM DOORS AND FRAMES
08130 STAINLESS STEEL DOORS AND FRAMES
08140 BRONZE DOORS AND FRAMES
08200 WOOD AND PLASTIC DOORS
08210 WOOD DOOR
08220 PLASTIC DOOR
08300 SPECIAL DOORS
08305 ACCESS DOOR
08310 SLIDING METAL FIRE DOOR
08320 METAL-CLAD DOOR
08330 COILING DOOR
08340 COILING GRILLE
08350 FOLDING DOOR
08360 OVERHEAD DOOR
08370 SLIDING GLASS DOOR
08380 SOUND-RETARDANT DOOR
08390 SCREEN DOOR
08391 STORM DOOR
08400 ENTRANCES AND STOREFRONTS
08410 ALUMINUM STOREFRONT
08420 ENTRANCE DOOR
08425 AUTOMATIC ENTRANCE DOOR
08450 REVOLVING DOOR
08500 METAL WINDOWS
08510 STEEL WINDOWS
08511 AWNING WINDOW
08512 CASEMENT
08513 DOUBLE-HUNG
08514 JALOUSIE
08515 PIVOTED
08516 PROJECTED
08520 ALUMINUM WINDOWS
08521 AWNING WINDOW
08522 CASEMENT
08523 DOUBLE-HUNG
08524 JALOUSIE

08525 PIVOTED
08526 PROJECTED
08527 SLIDING
08600 WOOD AND PLASTIC WINDOWS
08610 WOOD WINDOW
08611 AWNING WINDOW
08612 CASEMENT
08613 DOUBLE-HUNG
08614 JALOUSIE
08615 PIVOTED
08616 PROJECTED
08617 SLIDING
08650 SPECIAL WINDOWS
08651 SECURITY WINDOW
08700 HARDWARE
08721 AUTOMATIC-DOOR EQUIPMENT
08725 WINDOW OPERATOR
08730 WEATHERSTRIPPING
08740 THRESHOLD
08800 GLAZING
08810 GLASS
08830 MIRROR GLASS
08840 GLAZING PLASTIC
08900 GLAZED CURTAIN WALLS
08911 GLAZED STEEL CURTAIN WALL
08912 GLAZED ALUMINUM CURTAIN WALL
08913 GLAZED STAINLESS STEEL CURTAIN WALL
08914 GLAZED BRONZE CURTAIN WALL
08915 GLAZED WOOD CURTAIN WALL
08920 TRANSLUCENT WALL AND SKYLIGHT SYSTEM

DIVISION 9 FINISHES

09100 METAL SUPPORT SYSTEMS
09110 NON-LOAD-BEARING WALL FRAME
09111 FRAME AT FLOOR
09112 CONNECTION AT BEARING WALL
09113 CONNECTI   AT COLUMN
09114 INTERSECTION WITH PARTITION
09120 SUSPENDED CEILING SYSTEM
    Carrying Channel
    Expansion Joint
    Ceiling at Wall
09121 ANCHOR FOR FIXTURES/EQUIPMENT
    Bracing
    Hanger
09122 COFFER
    Light Pocket
    Recessed Light
09123 CONTROL JOINT
    Expansion Joint
09124 COVE
    Valance
09125 SOFFIT
    Dropped Ceiling
09126 TRACK

    Track Pocket
09127 THRU-CEILING OPENING
09130 ACOUSTICAL SUSPENSION SYSTEM
09200 LATH AND PLASTER
09201 FURRING AND LATHING
09210 GYPSUM PLASTER
09211 GYPSUM LATH AND PLASTER WALLS
    Anchor for Fixtures/Equipment
    Base
    At Corner (Corner Guard, see 10260—Wall Guards)
    At Ceiling
    At Existing Construction
    At Intersecting Wall
    At Slab Above
    Control Joint
    Thru-Wall Opening
09212 GYPSUM LATH AND PLASTER CEILING
    Anchor for Fixtures/Equipment
    Coffer
        Lighting Pocket
        Recessed Light
    Control Joint
    Cove
        Valance
    Soffit
        Dropped Ceiling
        Exterior Soffit
    Track
        Track Pocket
    Thru-Ceiling Opening
09215 VENEER PLASTER
09216 METAL LATH AND PLASTER WALLS
    Anchor for Fixtures/Equipment
    Coffer
        Lighting Pocket
        Recessed Light
    Control Joint
    Cove
        Valance
    Soffit
        Dropped Ceiling
        Exterior Soffit
    Track
        Track Pocket
    Thru-Ceiling Opening
09217 METAL LATH AND PLASTER CEILING
    Anchor for Fixtures/Equipment
    Coffer
        Lighting Pocket
        Recessed Light
    Control Joint
    Cove
        Valance
    Soffit
        Dropped Ceiling
        Exterior Soffit
    Track
        Track Pocket

Thru-Ceiling Opening
09220 PORTLAND CEMENT PLASTER
09225 ADOBE FINISH
09230 AGGREGATE COATINGS
09250 GYPSUM BOARD (DRYWALL)
09260 GYPSUM WALLBOARD SYSTEMS
09261 GYPSUM BOARD (DRYWALL) ON METAL FRAMING, ONE-HOUR CONSTRUCTION
    Anchor for Fixtures/Equipment
    Base
        Connection at Floor
    At Corner (Corner Guard, see 10260—Wall Guards)
        90° Corner
        Angle Corner
    At Ceiling
    At Existing Construction
    At Intersecting Wall
        Three-Way Intersection
        Four-Way Intersection
    At Slab Above
    Stub Wall
    Thru-Wall Opening
        Pass-Thru
        Sill/Jamb/Head
        Sleeve
09262 GYPSUM BOARD (DRYWALL) ON METAL FRAMING, TWO-HOUR CONSTRUCTION
    Anchor for Fixtures/Equipment
    Base
        Connection at Floor
    At Corner (Corner Guard, see 10260—Wall Guards)
        90° Corner
        Angle Corner
    At Ceiling
    At Existing Construction
    At Intersecting Wall
        Three-Way Intersection
        Four-Way Intersection
    At Slab Above
    Stub Wall
    Thru-Wall Opening
        Pass-Thru
        Sill/Jamb/Head
        Sleeve
09263 GYPSUM BOARD (DRYWALL) ON WOOD FRAMING, ONE-HOUR CONSTRUCTION
    Anchor for Fixtures/Equipment
    Base
    Connection at Floor
    At Corner (Corner Guard, see 10260—Wall Guards)
        90° Corner
        Angle Corner
    At Ceiling
    At Existing Construction
    At Intersecting Wall
        Three-Way Intersection

Four-Way Intersection
At Slab Above
Stub Wall
Thru-Wall Opening
    Pass-Thru
    Sill/Jamb/Head
    Sleeve
09264 GYPSUM BOARD (DRYWALL) ON WOOD FRAMING, TWO-HOUR CONSTRUCTION
    Anchor for Fixtures/Equipment
    Base
    Connection at Floor
    At Corner (Corner Guard, see 10260—Wall Guards)
        90° Corner
        Angle Corner
    At Ceiling
    At Existing Construction
    At Intersecting Wall
        Three-Way Intersection
        Four-Way Intersection
    At Slab Above
    Stub Wall
    Thru-Wall Opening
        Pass-Thru
        Sill/Jamb/Head
        Sleeve
09265 SOLID GYPSUM BOARD (DRYWALL) PARTITIONS, ONE-HOUR CONSTRUCTION
    Anchor for Fixtures
    Base
    At Corner (Corner Guard, see 10260—Wall Guards)
        90° Corner
        Angle Corner
    At Ceiling
    At Existing Construction
    At Intersecting Wall
        Three-Way Intersection
        Four-Way Intersection
    At Slab Above
    Stub Wall
    Thru-Wall Opening
        Pass-Thru
        Sill/Jamb/Head
        Sleeve
09266 SOLID GYPSUM BOARD (DRYWALL) PARTITIONS, TWO-HOUR CONSTRUCTION
    Anchor for Fixtures
    Base
    At Corner (Corner Guard, see 10260—Wall Guards)
        90° Corner
        Angle Corner
    At Ceiling
    At Existing Construction
    At Intersecting Wall
        Three-Way Intersection
        Four-Way Intersection

At Slab Above
Stub Wall
Thru-Wall Opening
  Pass-Thru
  Sill/Jamb/Head
  Sleeve
09267 SOLID GYPSUM BOARD (DRYWALL) PARTITIONS, THREE-HOUR CONSTRUCTION
  Anchor for Fixtures
  Base
  At Corner (Corner Guard, see 10260—Wall Guards)
    90° Corner
    Angle Corner
  At Ceiling
  At Existing Construction
  At Intersecting Wall
    Three-Way Intersection
    Four-Way Intersection
  At Slab Above
  Stub Wall
  Thru-Wall Opening
    Pass-Thru
    Sill/Jamb/Head
    Sleeve
09268 SOLID GYPSUM BOARD (DRYWALL) PARTITIONS, FOUR-HOUR CONSTRUCTION
  Anchor for Fixtures
  Base
  At Corner (Corner Guard, see 10260—Wall Guards)
    90° Corner
    Angle Corner
  At Ceiling
  At Existing Construction
  At Intersecting Wall
    Three-Way Intersection
    Four-Way Intersection
  At Slab Above
  Stub Wall
  Thru-Wall Opening
    Pass-Thru
    Sill/Jamb/Head
    Sleeve
09270 GYPSUM WALLBOARD (DRYWALL) CEILING
09271 ANCHOR FOR FIXTURES/EQUIPMENT
09272 COFFER
  Lighting Pocket
  Recessed Light
09273 CONTROL JOINT
  Expansion Joint
09274 COVE
  Valance
09275 SOFFIT
  Dropped Ceiling
  Exterior Soffit
09276 TRACK
  Track Pocket

09277 THRU-CEILING OPENING
09300 TILE
09310 CERAMIC TILE
  Floor
  Base
  Wainscot
  Wall
  Shower
  Tub
  Drain
  Counter top
09330 QUARRY TILE
  Floor
  Base
  Wainscot
  Wall
  Counter top
09332 SLATE TILE
  Floor
  Base
  Wainscot
  Wall
09340 MARBLE TILE
  Floor
  Base
  Wainscot
  Wall
  Counter top
09380 CONDUCTIVE TILE
  Floor
  Base
  Wainscot
  Wall
09400 TERRAZZO
09410 PORTLAND CEMENT TERRAZZO
09420 PRECAST TERRAZZO
09430 CONDUCTIVE TERRAZZO
09500 ACOUSTICAL TREATMENT
09510 ACOUSTICAL CEILING (also see 09130—Acoustical Suspension System)
09520 ACOUSTICAL WALL
09530 ACOUSTICAL INSULATION
09550 WOOD FLOOR
09551 WOOD BASE
09560 WOOD STRIP FLOOR
09570 WOOD PARQUET FLOOR
09580 PLYWOOD BLOCK FLOOR
09590 RESILIENT WOOD FLOOR
09595 WOOD BLOCK INDUSTRIAL FLOOR
09600 STONE AND BRICK FLOOR
09610 STONE FLOOR
09611 FLAGSTONE FLOOR
09612 SLATE FLOOR
09613 MARBLE FLOOR
09614 GRANITE FLOOR
09620 BRICK FLOOR
09650 RESILIENT FLOOR
09660 RESILIENT TILE FLOOR
09680 CARPETING

09700 SPECIAL FLOORING
09701 RESINOUS FLOOR
09710 MAGNESIUM OXYCHLORIDE
09720 EPOXY-MARBLE CHIP
09730 ELASTOMERIC-LIQUID FLOOR
09740 HEAVY-DUTY CONCRETE TOPPINGS
09750 MASTIC FILLS
09760 FLOOR TREATMENT
09800 SPECIAL COATING
09840 FIRE-RESISTANT COATING
09845 INTUMESCENT COATING
09950 WALL COVERING
09970 PREFINISHED PANEL

DIVISION 10
SPECIALTIES

10100 CHALKBOARD/TACKBOARD
10110 CHALKBOARD
10120 TACKBOARD
10150 COMPARTMENTS AND CUBICLES
10151 HOSPITAL CUBICLE
10160 TOILET PARTITION/URINAL SCREEN
10161 LAMINATED PLASTIC TOILET PARTITION/ URINAL SCREEN
10162 METAL TOILET PARTITION/URINAL SCREEN
10163 STONE PARTITION
10170 SHOWER COMPARTMENT
10171 DRESSING COMPARTMENT
10200 LOUVERS AND VENTS
10201 DOOR LOUVER
10202 WALL LOUVER
10220 ACCESS PANELS
10221 FLOOR PANEL
10222 WALL PANEL
10223 CEILING PANEL
10240 GRILLE AND SCREEN
  Ceiling
  Floor
  Wall
10250 SERVICE WALL
10260 WALL GUARDS
10265 CORNER GUARDS
10270 ACCESS FLOORING
10290 PEST CONTROL
10300 FIREPLACES AND STOVES
10301 PREFABRICATED FIREPLACE
10340 PREFABRICATED STEEPLES, SPIRES, AND CUPOLAS
10350 FLAGPOLES
10400 IDENTIFYING DEVICES
10410 DIRECTORY
10415 BULLETIN BOARD
10420 PLAQUE
10430 ILLUMINATED SIGNS
10431 FIRE EXIT SIGN
10440 SIGN

10450 PEDESTRIAN CONTROL DEVICES
10500 LOCKERS
  Locker Base
10520 FIRE EXTINGUISHERS, CABINETS, AND ACCESSORIES
10530 PROTECTIVE COVERS
10531 WALKWAY COVER
10532 CAR SHELTER
10535 AWNING
10550 POSTAL SPECIALTIES
10551 MAIL CHUTE
10552 MAIL BOX
10600 PARTITIONS
10601 MESH PARTITION
10610 DEMOUNTABLE PARTITION
10620 FOLDING PARTITION
10650 SCALES
10670 STORAGE SHELVING
10671 METAL STORAGE SHELVING
10700 SUN-CONTROL DEVICES (EXTERIOR)
10750 TELEPHONE ENCLOSURES
10751 TELEPHONE BOOTH
10752 TELEPHONE DIRECTORY UNIT
10753 TELEPHONE SHELVES
10800 TOILET AND BATH ACCESSORIES
10810 METAL-FRAMED MIRROR
10811 RECESSED PAPER TOWEL HOLDER
10812 RECESSED WASTE PAPER RECEPTACLE
10815 HANDICAPPED RAIL
10900 WARDROBE SPECIALTIES

DIVISION 11
EQUIPMENT

11010 MAINTENANCE EQUIPMENT
11011 VACUUM CLEANING EQUIPMENT
11012 WINDOW-WASHING EQUIPMENT
11020 SECURITY AND VAULT EQUIPMENT
11021 VAULT DOOR/DAY GATE
11022 TELLER WINDOW UNIT
  Service Window Unit
11025 AUTOMATIC-BANKING EQUIPMENT
11026 DEPOSITORY
11028 SAFE
11029 SAFETY DEPOSIT BOXES
11030 CHECKROOM EQUIPMENT
11040 ECCLESIASTICAL EQUIPMENT
11050 LIBRARY EQUIPMENT
11051 BOOK-THEFT PROTECTION EQUIPMENT
11053 LIBRARY STACK
11060 THEATER AND STAGE EQUIPMENT
11061 RIGGING SYSTEM AND CONTROL
11062 STAGE CURTAIN
11063 STAGE LIFT SYSTEM
11064 LIGHTING SYSTEM AND CONTROL
11065 ACOUSTICAL SHELL SYSTEM
11070 MUSICAL EQUIPMENT
11071 ORGAN

197

DIVISION 12 FURNISHINGS

DIVISION 13 SPECIAL CONSTRUCTION

DIVISION 14
CONVEYING SYSTEMS

THE DETAIL CATALOG DIVISION INDEX

INSTRUCTIONS FOR USING THE DETAIL CATALOG DIVISION INDEX

The detail catalog is divided according to the normal phases and sequence of construction:

Section One—Site Work Details

Section Two—Structural Details

Section Three—Roofing Construction Details

Section Four—Exterior Enclosure Details

Section Five—Interior Construction Details

Subsections are mainly titled according to the commonly used CSI Masterformat system, but are listed alphabetically instead of numerically. Thus, within Section One—Site Work Details, the subheadings follow alphabetically as Drainage, Earthwork, Equipment, etc. A list of "suggested subdivisions" is included with each general subheading. These, too, are listed alphabetically.

The detail catalog can follow any format or sequence your office finds convenient. It's strictly for in-house use, so feel free to modify its index any way you choose. The only rule is that you use the CSI Masterformat numbers consistently for your details.

DETAIL CATALOG DIVISIONS

SECTION ONE—SITE WORK DETAILS

DRAINAGE (02400)
Suggested Subdivisions:
   Area Drains (02420)
   Catch Basins (02431)
   Culverts (02434)
   Curb Inlets (02432)
   Foundation Drains (02411)
   Splash Blocks (02435)
   Storm Drains (02420)
   Subdrains (02410)
EARTHWORK (02200)
EQUIPMENT, FIXTURES, AND FURNISHINGS (SITE IMPROVEMENTS) (02440)
Suggested Subdivisions:
   Bicycle Racks (02457)
   Fencing (02444–02446)
   Furniture (Benches, Seats) (02471)
   Guardrails (Bollards, Stanchions) (02451)
   Parking (02455–02456)
   Play Equipment (02460, 02463)
   Shelters (02477)
   Signs and Signals (02452, 02453)
LANDSCAPING (02480)
Suggested Subdivisions:
   Aggregate/Chip Beds (02495, 02496)
   Shrub/Tree Planting (02491, 02492)
   Planters (02447)
PAVING AND SURFACING (02500)
Suggested Subdivisions:
   Asphalt (02516)

Bark Surfacing (02519)
Brick (02514)
Concrete (Drives, Roads) (02515)
Concrete (Walks, Steps) (02529)
Curbs (Asphalt, Concrete, Granite) (02525–02528)
Gravel Surfacing (02529)
Marking (Crosswalk, Direction, Stripes) (02577)
Sports Paving (02530)
Stone (02517)
PILES (Caissons, Cofferdams) (02350–02380)
PIPED UTILITIES (02600–02700)
Suggested Subdivisions:
Distribution Boxes (02745)
Grease Interceptors (02742)
Hydrants (02644)
Manholes (02601)
Sanitary Sewerage (02722)
Septic Tanks (02743)
Storm Sewerage (02721)
Water Wells (02730)
POWER AND COMMUNICATION (02800)
SITE IMPROVEMENTS (See Equipment, Fixtures, and Furnishings) (02440)

SECTION TWO— STRUCTURAL DETAILS

CONCRETE (03000)
Suggested Subdivisions:
Accessories (03250)
Movement Joints (03251)
Concrete, Cast-In-Place (03300)
Concrete Beams (03318)
Concrete Columns (03317)
Concrete Retaining Walls (03302)
Cantilever Retaining Wall
Gravity Retaining Wall
L-Type
T-Type
Weep Hole
Concrete Spandrel Beams (03319)
Concrete Walls (03316)
Construction Joint
Isolation Joint
Movement Joint (Expansion-Contraction)
Flat Plate Slabs (03323)
Flat Solid Concrete Slabs (03321)
Footings (03305)
Spread Footing
Step Footing
Foundation Walls (Heavy-Frame Construction) (03307)
Foundation Walls (Light-Frame Construction) (03306)
Grade Beams (03304)
Tie Beam
Metal Deck Concrete Slabs (03325)
One-Way Concrete Joist Slabs (03327)

Piers (03303)
Bell Pier
Socket Pier
Slab On Grade (Heavy-Frame Construction) (03309)
Construction Joint
Control Joint
Depressed Slab
Isolation Joint
Raised Slab
Slab on Grade (Light-Frame Construction) (03308)
Construction Joint
Control Joint
Depressed Slab
Isolation Joint
Raised Slab
Waffle Slabs (03329)
Concrete, Precast (03400)
Precast Beams (03416)
Precast Concrete Decking (03412)
Precast Concrete Plank (03413)
Hollow-Core Slab
Solid Flat Slab
Precast Concrete Wall Panels (03411)
Precast Girders (03415)
Precast Joists (03417)
Prestressed, Precast Concrete Sections (03420)
Stemmed Deck (03414)
Double T
Single T
Tilt-Up Concrete (03430)
MASONRY (04000)
Suggested Subdivisions:
Accessories (04150)
Anchors (04170)
Control Joints (04180)
Joint Reinforcement (04160)
Unit Masonry (04200)
Adobe (04212)
Anchors To Framing (04208)
Anchor to Concrete
Anchor to Steel
Anchor to Wood
Brick (04210)
Cavity Wall Lintels (04203)
Cavity Wall at Roof (04206)
Shelf Angle
Cavity Wall at Spandrel (04205)
Shelf Angle
Concrete Unit Masonry (04220)
Reinforced Concrete Block Foundation Walls (04229)
Reinforced Concrete Block Walls (04230)
Retaining Wall, Concrete Block (04228)
METALS (05000)
Suggested Subdivisions:
Metal Decking (05300)
Metal Joists (05200)

Aluminum Joists (05220)
Steel Joists (05210)
Joist at Wall
Joist at Beam
Structural Metal Framing (05100)
Framing Systems (05160)
Structural Aluminum (05130)
Structural Steel (05120)
Tubular Steel (05122)
Tubular Steel Columns (05123)
Base Plate
Tubular Steel Column To Beam (05124)
WF Column (05110)
Base Plate
WF Column To Steel Beam (05111)
WOOD AND PLASTIC (06000)
Suggested Subdivisions:
Fasteners and Supports (06050)
Heavy-Timber Construction (06130)
Mill-Framed Structures (06132)
Pole Construction (06133)
Prefabricated Structural Wood (06170)
Glu-Lam (06180)
Wood Trusses (06190)
Rough Carpentry (06100)
Beam Splice (06105)
Framing, Heavy Construction (06110)
Framing, Light Construction (06111)
Girder and Beam (06106)
Post at Beam (06104)
Post at Base (06102)
Post at Grade (06101)
Post at Header (06103)
Preassembled Components (06112)
Sheathing (06113)
Structural Plywood (06114)
Wood-Metal Systems (06150)
Wood Chord Metal Joists (06151)

SECTION THREE—ROOF CONSTRUCTION DETAILS

CEMENTITIOUS DECKS (03500)
Suggested Subdivisions:
Asphalt and Perlite Concrete Deck (03550)
Cementitious Wood Fiber Systems (03530)
Composite Concrete and Insulation Deck (03540)
Gypsum Concrete Deck (03510)
Precast Gypsum Plank (03515)
CONCRETE ACCESSORIES (03250)
Suggested Subdivisions:
Anchors (03252)
Expansion-Contraction Joints (03251)
Water Stops (03253)
CONCRETE, CAST-IN-PLACE (03300)
Suggested Subdivisions:
Flat Plate Slab (03323)

Flat Solid Concrete Slab (03321)
Metal Deck Concrete Slab (03325)
One-Way Concrete Joist Slab (03327)
Waffle Slab (03329)
CONCRETE, PRECAST (03400)
Suggested Subdivisions:
Precast Concrete Deck (03412)
Precast Concrete Plank (03413)
Stemmed Deck (03414)
CONVEYING SYSTEMS (Elevator, Elevator Penthouse) (14000)
EQUIPMENT (Maintenance, Window-Washing Systems) (11000)
METALS (05000)
Suggested Subdivisions:
Decking (05300)
Metal Roof Deck (05310)
Expansion Joints (05800)
Exterior Expansion Joint (05802)
Joists (05200)
Aluminum Joist (05220)
Steel Joist (05210)
Metal Fabrications (05500)
Exterior Fire Escape (05514)
Ladder Safety Cage (05519)
Post-and-Pipe Railing (05520)
Roof Ladder (05517)
Ship's Ladder (05516)
Stair (05510)
Tower Ladder (05518)
SPECIAL CONSTRUCTION (13000)
Suggested Subdivisions:
Solar Energy Systems (13980)
Packaged Solar System (13987)
Solar Flat Plate Collectors (13981)
SPECIALTIES (10000)
Suggested Subdivisions:
Flagpoles (10350)
Prefabricated Steeples, Spires, and Cupolas (10340)
THERMAL AND MOISTURE PROTECTION (07000)
Suggested Subdivisions:
Flashing and Sheet Metal (07600)
Chimney Flashing (07608)
Curb Flashing (07607)
Downspout (07632)
Flashing at Roof Edge (07605)
Gravel Stop (07660)
Gutter (07631)
Metal Coping (07665)
Metal Fascia (07604)
Parapet Flashing (07603)
Roofing Specialties (07630)
Scupper (07633)
Sheet Metal Flashing (07620)
Sheet Metal Trim
Sheet Metal Roofing (07610)
Stack Flashing (07606)
Insulation (07200)
Roof and Deck Insulation (07220)

Membrane Roofing (07500)
  Built-Up Bituminous Roofing (07510)
  Elastic Sheet Roofing (07530)
  Prepared Roll Roofing (07520)
Preformed Roofing and Siding (07400)
Roof Accessories (07800)
  Breather Vent (07825)
  Curb, Mechanical Equipment (07824)
  Curb, Prefab (07850)
  Curb, Skylight (07810)
  Drain (07822)
  Flagpole Support (07874)
  Gravity Ventilator (07840)
  Guy Wire Connection (07871)
  Light Support (07873)
  Pitch Pocket (07870)
  Prefab Expansion Joint (07860)
  Roof Deck Drain (07823)
  Roof Hatch (07830)
  Roof Ladder Support (07875)
  Roof Railing Support (07876)
  Sign Support (07872)
  Skylight (07810)
  Stack (07820)
  Vent Stack (07821)
Shingles and Roofing Tiles (07300)
  Roofing Tile (07320)
  Shingles (07310)
Traffic Topping (07570)
WOOD (06000)
Suggested Subdivisions:
  Prefabricated Structural Wood (06170)
    Glu-Lam (06180)
    Wood Trusses (06190)
  Rough Carpentry (06100)
    Wall at Roof (06119)
    Wood Decking (06125)

SECTION FOUR—EXTERIOR ENCLOSURE DETAILS

CONCRETE (03000)
Suggested Subdivisions:
  Concrete, Cast-in-Place (03300)
    Architectural Concrete (Ornamental) (03330)
    Concrete Columns (03317)
    Concrete Walls (03316)
  Concrete, Precast (03400)
    Architectural Precast Concrete (03450)
    Precast Concrete Wall Panels (03411)
    Tilt-Up Concrete (03430)
  Special Concrete Finishes (03350)
DOORS (08000)
Suggested Subdivisions:
  Entrances and Storefronts (08400)
    Aluminum Storefront (08410)
    Automatic Entrance Door (08425)

Entrance Door (08420)
    Revolving Door (08450)
Metal Doors and Frames (08100)
  Aluminum Doors and Frames (08120)
  Steel Doors (08110–08112)
  Steel Frames (08111–08113)
Special Doors (08300)
  Access Door (08305)
  Coiling Door (08330)
  Coiling Grille (08340)
  Folding Door (08350)
  Metal-Clad Door (08320)
  Overhead Door (08360)
  Screen Door (08390)
  Sliding Glass Door (08370)
  Sliding Metal Fire Door (08310)
  Sound-Retardant Door (08380)
Wood and Plastic Doors (08200)
EQUIPMENT (11000)
Suggested Subdivisions:
  Display Case (Exterior) (11101)
  Loading Dock Equipment (11160)
    Dock Bumper (11165)
  Maintenance Equipment (11010)
    Window-Washing Equipment (11012)
  Security and Vault Equipment (11020)
    Automatic-Banking Equipment (11025)
    Depository (11026)
    Teller Window Unit (11022)
  Waste-Handling Equipment (11170)
FINISHES (09000)
Suggested Subdivisions:
  Aggregate Coatings (09230)
  Lath and Plaster (Stucco) (09200)
  Tile (09300)
    Ceramic Tile (09310)
    Marble Tile (09340)
    Quarry Tile (09330)
    Slate Tile (09332)
FURNISHINGS (12000)
Suggested Subdivisions:
  Artwork (12100)
MASONRY (04000)
Suggested Subdivisions:
  Accessories (04150)
    Anchor (04170)
    Control Joint (04180)
    Joint Reinforcement (04160)
  Stone (04400)
    Cut Stone (04420)
    Flagstone (04440)
    Marble Veneer (04451)
    Natural Stone Veneer (04450)
    Rough Stone (04410)
  Unit Masonry (04200)
    Adobe (04212)
    Anchor to Framing (04208)
      Anchor to Concrete
      Anchor to Steel

Anchor to Wood
Brick (04210)
Brick Cavity Wall (04214)
Brick Veneer (04215)
Cavity Wall at Parapet (04207)
Cavity Wall at Roof (04206)
Cavity Wall at Slab (04204)
Cavity Wall at Spandrel (04205)
Cavity Wall Lintel (04203)
Ceramic Veneer (04250)
Clay Backing Tile (04240)
Concrete Unit Masonry Cavity Wall (04220)
Exposed Aggregate Concrete Unit Masonry (04222)
Fluted Concrete Unit Masonry (04224)
Freestanding H.M.U. (Concrete Block) Wall (04226)
Glass Unit Masonry (04270)
Glazed Concrete Unit Masonry (04221)
Gypsum Unit Masonry (04280)
H.M.U. (Concrete Block) Retaining Wall (04228)
Interlocking Concrete Unit Masonry (04237)
Molded-Face Concrete Unit Masonry (04225)
Mortarless Concrete Unit Masonry (04238)
Preassembled Masonry Panels (04235)
Reinforced H.M.U. (Concrete Block) Wall (04230)
Split-Face Concrete Unit Masonry (04223)
Structural Glazed Tile Masonry (04213)
METALS (05000)
Suggested Subdivisions:
Cold-Formed Metal Framing (05400)
Metal Stud (Load-Bearing) (05410)
Expansion Joints (05800)
Exterior Expansion Joint (05802)
Metal Fabrications (05500)
Baluster and Railing (05523)
Exterior Fire Escape (05514)
Heating-Cooling Unit Enclosure (05551)
Metal Stair (05510–05513)
Ship's Ladder (05516)
Stair Handrail, Post-Mounted (05522)
Stair Handrail, Wall-Mounted (05521)
Wall Ladder (05515)
Ornamental Metal (05700)
Ornamental Handrail (05720)
Ornamental Sheet Metal (05730)
Ornamental Stair (05710)
SPECIAL CONSTRUCTION (13000)
Suggested Subdivisions:
Greenhouses (13123)
Solar Energy Systems (13980)
SPECIALTIES (10000)
Suggested Subdivisions:
Corner Guards (10265)
Grilles and Screens (10240)
Identifying Devices (10400)
Bulletin Board (10415)
Directory (10410)

Illuminated Signage (10430)
Plaque (10420)
Sign (10440)
Louvers and Vents (10200)
Door Louver (10201)
Wall Louver (10202)
Mail Box/Mail Chute (10550)
Pedestrian-Control Devices (10450)
Protective Covers (10530)
Awning (10535)
Car Shelter (10532)
Walkway Cover (10531)
Sun-Control Devices (Exterior) (10700)
Telephone Enclosures (10750)
Wall Guards (10260)
THERMAL AND MOISTURE PROTECTION (07000)
Suggested Subdivisions:
Dampproofing (07150)
Fireproofing (07250)
Shingles (07300)
Flashing and Sheet Metal (07600)
Cornice (07603)
Damp Course (07601)
Downspout (07632)
Parapet Flashing (07603)
Sheet Metal Flashing/Trim (07620)
Wall Flashing (07602)
Insulation (07200)
Building Insulation (07210)
Joint Sealants (07900)
Joint Filler (07910)
Sealant/Caulking (07920)
Metal Fascia (07604)
Preformed Siding (07400)
Composite Building Panels (07420)
Plywood Siding (07465)
Preformed Wall/Roof Panels (07410)
Shingles (07300)
Wood Siding (07461)
Waterproofing (07100)
WINDOWS (08000)
Suggested Subdivisions:
Glazed Curtain Walls (08900)
Aluminum Curtain Wall (08912)
Bronze Curtain Wall (08914)
Stainless Steel Curtain Wall (08913)
Steel Curtain Wall (08911)
Wood Curtain Wall (08915)
Translucent Wall and Skylight System (08920)
Glazing (08800)
Glass (08810)
Glazing Plastics (08840)
Hardware (08700)
Automatic-Door Equipment (08721)
Threshold (08740)
Weather Stripping (08730)
Window Operator (08725)
Metal Windows (08500)
Steel Window (08510)

Awning Window (08511)
Casement (08512)
Double-Hung (08513)
Jalousie (08514)
Pivoted (08515)
Projected (08516)
Aluminum Windows (08520)
Awning Window (08521)
Casement (08522)
Double-Hung (08523)
Jalousie (08524)
Pivoted (08525)
Projected (08526)
Sliding (08527)
Special Windows (08650)
Security Window (08651)
Wood and Plastic Windows (08600)
Wood Window (08610)
Awning Window (08611)
Casement (08612)
Double-Hung (08613)
Jalousie (08614)
Pivoted (08615)
Projected (08616)
Sliding (08617)
WOOD AND PLASTIC (06000)
Suggested Subdivisions:
Architectural Woodwork (06400)
Stair Work (06430)
Fasteners and Supports (06050)
Rough Carpentry (06100)
Framing, Heavy Construction (06110)
Framing, Light (06111)
Preassembled Components (06112)
Sheathing (06113)
Structural Plywood (06114)
Thru-Wall Opening (06117)
Wall at Ceiling (06118)
Wall at Floor (06116)
Wall at Foundation (06115)
Wall at Roof (06119)
Finish Carpentry (06200)
Exterior Stair Framing (06434)
Millwork (06220)
Jamb Trim
Wall Trim
Heavy-Timber Construction (06130)
Mill-Framed Structures (06132)

SECTION FIVE— INTERIOR CONSTRUCTION DETAILS

CONCRETE (03000)
Suggested Subdivisions:
Concrete, Cast-In-Place (03300)
Architectural Concrete (Ornamental) (03330)

Concrete Beams (03318)
Concrete Columns (03317)
Concrete Walls (03316)
Slabs (03321–03329)
Slab on Grade (Heavy-Frame Construction) (03309)
Slab on Grade (Light-Frame Construction) (03308)
Concrete, Precast (03400)
Architectural Precast Concrete (03450)
Architectural Precast Wall Panels (03451)
Precast Wall Panels (03411)
Special Concrete Finishes (03350)
CONVEYING SYSTEMS (14000)
Suggested Subdivisions:
Dumbwaiters (14100)
Elevators (14200)
Passenger Elevator (14210)
Freight Elevator (14220)
Hoists and Cranes (14300)
Lifts (14400)
Material-Handling Systems (14500)
Chute (14560)
Conveyor (14550)
Pneumatic Tube (14581)
Moving Stairs and Walks (14700)
Escalator (14710)
Moving Walk (14720)
Turntable (14600)
DOORS (08000)
Suggested Subdivisions:
Glazing (08800)
Glass (08810)
Glazing Plastics (08840)
Mirror Glass (08830)
Hardware (08700)
Automatic-Door Equipment (08721)
Metal Doors and Frames (08100)
Aluminum Doors and Frames (08120)
Bronze Doors and Frames (08140)
Stainless Steel Doors and Frames (08130)
Steel Door (08110, 08112, 08115)
Steel Frame (08111, 08112, 08115)
Special Doors (08300)
Access Door (08305)
Coiling Door (08330)
Coiling Grille (08340)
Folding Door (08350)
Metal-Clad Door (08320)
Overhead Door (08360)
Sliding Glass Door (08370)
Sliding Metal Fire Door (08310)
Sound-Retardant Door (08380)
Wood and Plastic Doors (08200)
Plastic Door (08220)
Wood Door (08210)
EQUIPMENT (11000)
Suggested Subdivisions:
Audiovisual Equipment (11130)

Checkroom Equipment (11030)
Darkroom Equipment (11470)
Detention Equipment (11190)
Food Service Equipment (11400)
Industrial and Process Equipment (11500)
Laboratory Equipment (11600)
Library Equipment (11050)
Maintenance Equipment (11010)
  Vacuum Cleaning Equipment (11011)
Medical Equipment (11700)
Mercantile Equipment (11100)
Mortuary Equipment (11780)
Musical Equipment (11070)
Residential Equipment (11450)
  Disappearing Stair (11454)
  Kitchen Equipment (11451)
  Laundry Equipment (11452)
Security and Vault Equipment (11020)
Telecommunication Equipment (11800)
Theater and Stage Equipment (11060)
Unit Kitchens and Cabinets (11460)
Vending Equipment (11120)
Waste-Handling Equipment (11170)
FINISHES (09000)
Suggested Subdivisions:
  Acoustical Treatment (09500)
    Acoustical Ceiling (09510) (also see 09130)
    Acoustical Insulation (09530)
    Acoustical Wall (09520)
  Aggregate Coatings (09230)
  Fire-Resistant Coating (09840)
  Floor, Stone, and Brick (09600)
  Floor, Wood (09550)
  Flooring, Special (09700)
  Gypsum Board (Drywall) (09250)
    On Metal Framing—One-Hour/Two-Hour (09261/09262)
    On Wood Framing—One-Hour/Two-Hour (09263/09264)
    Solid Gypsum Board Partitions—One-Hour (09265)
    Solid Gypsum Board Partitions—Two-Hour (09266)
    Solid Gypsum Board Partitions—Three-Hour (09267)
    Solid Gypsum Board Partitions—Four-Hour (09268)
  Lath and Plaster (09200)
    Gypsum Lath and Plaster Ceiling (09212)
    Gypsum Lath and Plaster Walls (09211)
    Metal Lath and Plaster Ceiling (09217)
    Metal Lath and Plaster Walls (09216)
  Metal Support Systems (09100)
    Acoustical Suspension System (09130)
    Wall Frame, Non-Load-Bearing (09110)
  Suspended Ceiling (09120)
  Terrazzo (09400)
  Tile (09300)
    Ceramic Tile (09310)

Conductive Tile (09380)
Marble Tile (09340)
Quarry Tile (09330)
Slate Tile (09332)
FURNISHINGS (12000)
Suggested Subdivisions:
  Artwork (12100)
  Cabinets and Casework, Manufactured (12300)
    Bank Fixtures and Casework (12310)
    Built-In Tables (12303)
    Display Casework (12380)
    Educational Cabinets and Casework (12325)
    Hospital Casework (12350)
    Hotel and Motel Casework (12370)
    Library Casework (12315)
    Medical and Laboratory Casework (12335)
    Metal Casework (12301)
    Residential Casework (12390)
    Restaurant and Cafeteria Fixtures and Casework (12320)
    Wood Casework (12302)
  Furniture and Accessories (12600)
  Multiple Seating (also see 13125, Grandstand) (12700)
    Auditorium and Theater Seating (12710)
    Bench (12771)
    Booth (12740)
    Multiple-Use Fixed Seating (12750)
    Pew (12770)
    Stadium and Arena Seating (12730)
    Telescoping Bleachers (12760)
  Plants and Planting, Interior (12800)
    Planter (12815)
  Rugs and Mats (12670)
    Foot Grille (12672)
    Mat Frame (12673)
  Window Treatment (12500)
    Drape and Curtain (12501)
MASONRY (04000)
Suggested Subdivisions:
  Accessories (04150)
    Anchors (04170)
    Control Joints (04180)
    Joint Reinforcement (04160)
  Stone (04400)
    Cut Stone (04420)
    Flagstone (04440)
    Marble Veneer (04451)
    Natural Stone Veneer (04450)
    Rough Stone (04410)
  Unit Masonry (04200)
    Adobe (04212)
    Anchor to Framing (04208)
    Brick (04210)
    Brick Veneer (04215)
    Ceramic Veneer (04250)
    Exposed Aggregate Concrete Unit Masonry (04222)
    Fluted Concrete Unit Masonry (04224)

Freestanding Brick Wall (04216)
Freestanding H.M.U. (Concrete Block) Wall (04226)
Glass Unit Masonry (04270)
Glazed Concrete Unit Masonry (04221)
Gypsum Unit Masonry (04280)
Interior Brick Partition (04217)
Interior H.M.U. (Concrete Block) Partition (04227)
Interlocking Concrete Unit Masonry (04237)
Molded-Face Concrete Unit Masonry (04225)
Mortarless Concrete Unit Masonry (04238)
Preassembled Masonry Panel (04235)
Sound-Absorbing Unit Masonry (04285)
Split-Face Concrete Unit Masonry (04223)
METALS (05000)
Suggested Subdivisions:
Cold-Formed Metal Framing (05400)
Cold-Formed Metal Joist (05420)
Metal Stud (Load-Bearing) (05410)
Expansion Joints (05800)
Interior Expansion Joint (05801)
Metal Fabrication (05500)
Baluster and Railing (05523)
Floor Plate (05531)
Grating (05530)
Heating-Cooling Unit Enclosure (05551)
Ladder, Wall (05515)
Metal Stair, Channel Stringer (05510)
Metal Stair, Open Riser (05512)
Metal Stair, Pan-type (05511)
Metal Tread Nosing (05513)
Post-and-Pipe Railing (05520)
Ship's Ladder (05516)
Stair Handrail, Post-Mounted (05522)
Stair Handrail, Wall-Mounted (05521)
Ornamental Metal (05700)
Ornamental Handrail (05720)
Ornamental Sheet Metal (05730)
Ornamental Stair (05710)
Prefabricated Spiral Stair (05715)
SPECIAL CONSTRUCTION (13000)
Suggested Subdivisions:
Ceilings, Integrated (13070)
Pool (13150)
Aquaria (13152)
Hot Tub (13154)
Swimming Pool (13151)
Therapeutic and Massage Pools (13153)
Sound, Vibration Control (13080)
Vaults (13140)
SPECIALTIES (10000)
Suggested Subdivisions:
Access Flooring (10270)
Chalkboard/Tackboard (10100)
Compartments and Cubicles (10150)
Dressing Compartment (10171)
Hospital Cubicle (10151)

Shower Compartment (10170)
Toilet Partition (10160)
Urinal Screen (10162)
Corner Guards (10265)
Fire Extinguishers, Cabinets and Accessories (10520)
Fireplaces and Stoves (10300)
Grilles and Screens (10240)
Identifying Devices (10400)
Bulletin Board (10415)
Directory (10410)
Fire Exit Sign (10431)
Illuminated Signage (10430)
Plaque (10420)
Sign (10440)
Lockers (10500)
Louvers and Vents (10200)
Partitions (10600)
Postal Specialties (10550)
Service Wall (10250)
Storage Shelving (10670)
Telephone Enclosures (10750)
Toilet and Bath Accessories (10800)
Handicapped Rail (10815)
Wall Guards (10260)
Wardrobe Specialties (10900)
THERMAL PROTECTION (07000)
Suggested Subdivisions:
Dampproofing (07150)
Fireproofing (07250)
Insulation (07200)
Building Insulation (07210)
Waterproofing (07100)
WOOD AND PLASTIC (06000)
Suggested Subdivisions:
Architectural Woodwork (06400)
Cabinet Work (06410) (See FURNISHINGS: Cabinets)
Counter top (06411) (See FURNISHINGS: Cabinets)
Paneling (06420)
Railing, Wood (06431)
Shelving (06412)
Stair Work (06430)
Stringer (06433)
Wardrobe Storage (06413)
Fasteners and Supports (06050)
Finish Carpentry (06200)
Millwork (06220)
Heavy-Timber Construction (06130)
Prefabricated Structural Wood (06170)
Glu-Lam (06180)
Wood Trusses (06190)
Rough Carpentry (06100)

ARCHITECTURAL WORKING-DRAWING CHECKLIST I: COMMERCIAL, INSTITUTIONAL, AND OTHER HEAVY-FRAME CONSTRUCTION

INTRODUCTION: A CHECKLIST FOR USING THE CHECKLIST

The checklists that follow are a tremendous aid for planning, scheduling, budgeting, supervising, and checking architectural working-drawing production. Later, as more complete advanced systems are developed, you'll be able to integrate all job decisions with notation, keynotes, standard details, specifications, etc. For now, and within the limits of textbook format, you'll find this manual checklist format to be extremely valuable. I suggest you copy it, add to it, and revise it as you see fit for your individual needs.

Here are the steps for getting best use of the Architectural Working-Drawing Checklists (Appendixes B1 and B2):

____ First, scan through the entire list. Note additions, revisions, or deletions required to bring the list into line with your office standards and range of practice.

____ When you do schematic and design phases of a project, add to this list those special items that come up that relate only to that particular project.

____ When checking a client's property survey, compare it with the survey data portion of this checklist, to be sure all required data have been included. (Site visits are also required to verify completeness and accuracy of a survey, so use the survey section of this list as a predesign site inspection guide.)

____ During project planning, mark all job items to be included in the working drawings. The marked list will serve then as a complete production task list for drafting personnel.

____ Require drafting staff members to countermark required checklist items as they complete those items on the working drawings.

____ Use the list for reference when writing specifications. Verify that construction elements not on the working drawings are covered in specifications and vice versa.

____ During progress print phases, mark the required items that have been incorporated satisfactorily in the drawings.

____ At the conclusion of the job use the entire list as your final overall double check of completeness, accuracy, and coordination.

You can upgrade the list if you like by tacking on special elements to your copy of the checklist. For example, you can add a code system to record reasons for decisions, when they were made, and by whom. Then the checklist becomes a job diary and project documentation tool similar to what I described earlier in the book. Or you can add in key number references to coordinate this list with your standard detail file system and keynote file system. I'm preparing such elaborations now which will be suited for both manual use and word processor or computer use. But meanwhile, there's no reason why you can't move ahead with the idea on your own.

COVER SHEET, SHEET FORMAT, AND TITLE BLOCKS

COVER SHEET/INDEX SHEET

- ☐ PROJECT NAME____ OWNER'S NAME____.
- ☐ PERSPECTIVE RENDERING OR PHOTO OF MODEL.
- ☐ INDEX OF ARCHITECTURAL AND CONSULTANTS' DRAWINGS.
- ☐ GENERAL NOTES.
- ☐ LEXICON OF ABBREVIATIONS AND NOMENCLATURE.
- ☐ LEGEND OF STRUCTURAL, ELECTRICAL, PLUMBING, AND HVAC SYMBOLS AND CONVENTIONS (if not shown on other drawings).
- ☐ LEGEND OF SITE WORK AND SURVEY DRAWING CONVENTIONS AND SYMBOLS (if not shown on site plan).
- ☐ LEGEND OF CONSTRUCTION MATERIALS INDICATIONS AND DRAWING CONVENTIONS.
- ☐ LEGEND OF SYMBOLS: ELEVATION POINT____ LEVEL LINE____ REVISIONS____ COLUMN OR MODULE GRID KEY____ BUILDING CROSS SECTION OR PARTIAL SECTION KEY____ WALL SECTION KEY____ DETAIL KEY____ WINDOW NUMBER, KEY TO SCHEDULE____ DOOR NUMBER, KEY TO SCHEDULE____ ROOM OR AREA NUMBER, KEY TO FINISH SCHEDULE____ STAIR KEY____ EQUIPMENT KEY____.

WORKING-DRAWING SHEET FORMAT

- ☐ SCHEDULE ON LEFT-HAND TRIM MARGIN TO RECORD DATE AND TIMES OF WORK PERIODS ON SHEET BY INDIVIDUAL STAFF MEMBERS.
- ☐ RECOMMENDED MAXIMUM DRAWING SIZE (to meet limits of "Scan" system): 30" × 40"____ MAIN BODY OF DRAWING SIZED IN 8½" × 11" INCREMENTS____.
- ☐ TICK MARKS PRINTED AT BORDER EDGES TO GUIDE FOLDING OF DRAWINGS INTO 8½" × 11" PACKETS.
- ☐ TRIM SPACE OF ½" OR MORE AROUND OUTER BORDER OF SHEET. (Permits elimination of bent corners, rips, and tears at edges by trimming sheet to final size prior to final printing.)
- ☐ NOTES AND GENERAL NOTES AT RIGHT-HAND SIDE OF TRACINGS.

- ☐ TITLE BLOCK AT LOWER RIGHT OR ACROSS RIGHT-HAND EDGE OF TRACING____ SHEET NUMBERS PROMINENT AT BOTTOM RIGHT-HAND CORNER____.

TITLE BLOCKS

- ☐ ARCHITECT'S NAME____ ADDRESS____ PHONE NUMBER____ REGISTRATION NUMBER OR OFFICIAL STAMP____.
- ☐ PROJECT NAME AND ADDRESS.
- ☐ OWNER'S NAME AND ADDRESS.
- ☐ CONSULTANTS: STRUCTURAL____ HVAC____ PLUMBING____ ELECTRICAL____ LIGHTING____ SOILS____ CIVIL ENGINEERING____ ACOUSTICAL____ AUDITORIUM____ KITCHEN AND FOOD SERVICE____ LANDSCAPING____ INTERIORS____.
- ☐ CONSULTANTS' NAMES____ ADDRESSES____ PHONE NUMBERS____ REGISTRATION NUMBERS OR STAMPS____.
- ☐ DRAWING TITLE AND SCALE.
- ☐ MINI OUTLINE PLAN OF BUILDING.
- ☐ PARTNER IN CHARGE____ CHIEF OR PROJECT ARCHITECT____ JOB OR TEAM CAPTAIN____.
- ☐ PROJECT MANAGER: NAME____ ADDRESS____ PHONE NUMBER____.
- ☐ DESIGNER AND DRAFTING TECHNICIANS. (A growing preference is to provide names of participants rather than initials.)
- ☐ INITIALS OR NAME OF DRAWING CHECKER.
- ☐ SPACE FOR REVISION DATES AND REVISION REFERENCE SYMBOLS.
- ☐ PROJECT NUMBER____ FILE NUMBER____.
- ☐ SPACE FOR GENERAL NOTES____ PREPRINTED GENERAL NOTES____.
- ☐ COPYRIGHT NOTICE OR NOTE ON RIGHTS AND RESTRICTIONS OF OWNERSHIP AND USE OF DRAWINGS.
- ☐ BUILDING AND PLANNING AUTHORITY NAMES____ ADDRESSES____ PHONE NUMBERS____ PERMIT NUMBERS____.
- ☐ SPACE FOR APPROVAL STAMPS OR INITIALS____ DATES OF APPROVALS____.
- ☐ CHECKING DATES____ JOB PHASE COMPLETION DATES____ CLIENT APPROVALS____.
- ☐ FINAL RELEASE DATE.
- ☐ DRAWING SHEET NUMBER AND TOTAL NUMBER OF DRAWINGS.
- ☐ SHEET NUMBER CODING FOR DRAWINGS OF DIFFERENT CONSULTANTS____ CODING FOR SEPARATE BUILDINGS OR PORTIONS OF PROJECT____.

SITE PLANS

Site plans may incorporate individual or combined sheets, including the following: survey, staking plan, test pit and test-boring plans, grading plan, drainage plan, demolition plan, excavation plan, landscaping plan, construction work site plan. Components of various possible plans are combined under appropriate subheadings in this checklist.

SURVEY

The survey may be incorporated "as is" within the working drawings or redrawn as an architectural drawing with additional construction data.

GENERAL REFERENCE DATA

☐ DRAWING TITLE AND SCALE (beneath drawing or on title block).

☐ ARROWS SHOWING COMPASS NORTH AND REFERENCE NORTH.

☐ PROPERTY ADDRESS: STREET____ CITY____ COUNTY____ STATE____.

☐ OWNER'S NAME____ ADDRESS____. (If not shown on title block.)

☐ PROPERTY LEGAL DESCRIPTION: LOT____ BLOCK____ TRACT____ SUBDIVISION____ MAP BOOK____.

☐ LOT AND BLOCK NUMBERS OF ADJACENT PROPERTIES.

☐ SURVEYOR'S NAME____ ADDRESS____ PHONE NUMBER____ REGISTRATION NUMBER____.

☐ SURVEY DATE.

PROPERTY DATA

☐ PROPERTY BOUNDARY LINES____ DIMENSIONS____ DIRECTIONS OF PROPERTY LINES SHOWN IN DEGREES, MINUTES, AND SECONDS____.

☐ MONUMENT____ BENCHMARK____ REFERENCE POINTS____ STAKES____ REFERENCE DATUM____.

☐ CONTOUR LINES WITH GRADE ELEVATIONS. (Contour lines aren't shown on flat lots, but grade elevation points are noted at corners and at sloping portions of the property.)

☐ ± NOTES ON ELEVATION POINTS IF GRADES ARE BOTH ABOVE AND BELOW DATUM POINT.

☐ ON-SITE AND ADJACENT EASEMENTS____ RIGHTS-OF-WAY____. (Identities and dimensions.)

☐ REQUIRED FRONT, REAR, AND SIDE SETBACK LINES____ DIMENSIONS TO PROPERTY LINE____.

☐ ADJACENT STREETS OR ROADS: NAMES____ WIDTH DIMENSIONS____ ELEVATION POINTS AT STREET CENTERLINES____ ELEVATION POINTS AT SIDES____ CONCRETE TRAFFIC ISLANDS____.

☐ ADJACENT STREETS OR ROADS: TOTAL RIGHTS-OF-WAY____ EASEMENTS FOR ROAD WIDENING____.

☐ STREET STORM MAINS____ BASINS____ DIRECTION AND ELEVATION POINTS OF DRAIN SLOPES____ INVERT ELEVATIONS____.

☐ STREET SEWER HOOKUP____ SEWER MANHOLE____ ELEVATION POINT AT TOP OF MANHOLE____ ELEVATION POINT AT SEWER INVERT____ DIRECTION OF FLOW____.

☐ STREET GAS AND ELECTRIC MANHOLES____ ELECTRICAL VAULTS____ OTHER UTILITIES IN STREET OR ADJACENT TO STREET____.

☐ SIDEWALK: LIFTS____ VAULTS____ VAULT SKYLIGHTS____ VAULT GRATINGS____ TUNNELS____ CHUTE AND FUEL PIPE INLETS____ METER BOXES____.

☐ EXISTING CURBS AND WALKWAYS ADJACENT TO PROPERTY____ DIMENSIONS AND ELEVATION POINTS____.

☐ CURB ELEVATION POINTS SHOWN AT EXTENSIONS OF PROPERTY LINES.

☐ TRAFFIC LIGHTS____ SIGNS____ PARKING METERS____ FIRE ALARM BOXES____ POLICE CALL BOXES____.

☐ FIRE HYDRANTS.

☐ EXISTING SIDEWALK TREES OR SHRUBS.

☐ EXISTING POWER POLE(S).

☐ GAS MAIN AND HOOKUP LOCATION.

☐ STREET WATER MAIN AND HOOKUP LOCATION____ EXISTING WATER METER LOCATION____.

☐ OVERHEAD OR BURIED CABLE ON, OR ADJACENT TO, PROPERTY.

☐ EXISTING BUILDINGS ON ADJACENT PROPERTY WITHIN 5' OF PROPERTY LINE____ KNOWN FUTURE ADJACENT CONSTRUCTION____.

☐ ADJACENT BUILDING FOOTINGS AND FOUNDATION WALLS AT PROPERTY LINE.

☐ EXISTING ON-SITE PAVING, CURBS, AND WALKWAYS: DIMENSIONS____ MATERIALS____ ELEVATION POINTS____.

☐ EXISTING ON-SITE STRUCTURES: IDENTITY____ SIZE____ FLOOR ELEVATION POINTS____.

☐ EXISTING FENCES AND WALLS: MATERI-

ALS____ HEIGHTS____.

☐ EXISTING UNDERGROUND STRUCTURES____ FOUNDATIONS____ HOLES AND TRENCHES____.

☐ EXISTING BURIED OR SURFACE TRASH.

☐ EXISTING ON-SITE UTILITIES____ SEWERS____ TUNNELS____ DRAINPIPES____.

☐ UTILITIES EASEMENTS.

☐ EXISTING ON-SITE OR ADJACENT DRAINAGE CULVERTS____ RIPRAP____.

☐ EXISTING SPRINGS OR WELLS.

☐ EXISTING PONDS____ CREEKS____ STREAMS____ OVERFLOW AREAS____.

☐ HIGH-WATER TABLE____ STANDING WATER____ MARSH____ QUICKSAND____.

☐ ROCK OUTCROPS____ ROCK ELEVATION POINTS____.

☐ EXISTING TREES: IDENTITIES____ TRUNK SIZES____ APPROXIMATE FOLIAGE AREA____. (Trees with trunks less than 6″ in diameter aren't usually noted. Architect may want smaller trees identified for removal and transplanting.)

☐ EXISTING MAJOR SHRUBS____ UNDERGROWTH____ GROUND COVER AREAS____.

SITE WORK

GENERAL REFERENCE DATA

☐ LEGAL SETBACK LINES.

☐ SITE PHOTOS (sometimes printed on site plan transparency print and number-keyed to arrows on site plan drawing).

☐ LEGENDS OF SITE PLAN SYMBOLS AND MATERIALS INDICATIONS.

☐ LANDSCAPE CONSULTANT'S NAME____ ADDRESS____ PHONE NUMBER____.

☐ CIVIL ENGINEER'S NAME____ ADDRESS____ PHONE NUMBER____ REGISTRATION NUMBER____.

☐ SOILS ENGINEER'S NAME____ ADDRESS____ PHONE NUMBER____ REGISTRATION NUMBER____.

☐ TEST-BORING CONTRACTOR'S NAME____ ADDRESS____ PHONE NUMBER____.

☐ SOIL TEST LAB'S NAME____ ADDRESS____ PHONE NUMBER____.

☐ SMALL-SCALE LOCATION OR VICINITY MAP SHOWING NEIGHBORING STREETS AND NEAREST MAJOR HIGHWAY ACCESS.

☐ NOTE REQUIRING BIDDERS TO VISIT SITE AND VERIFY CONDITIONS BEFORE SUBMITTING BIDS.

☐ PERCOLATION TEST PLAN____ SOILS TEST-BORING SCHEDULE AND PROFILE____.

☐ WORK NOT IN CONTRACT.

☐ DRAWING TITLE AND SCALE.

☐ ARROWS SHOWING COMPASS NORTH AND REFERENCE NORTH.

☐ PROPERTY SIZE IN SQUARE FEET OR ACRES.

☐ CONTRACT LIMIT LINES.

☐ LOCATION AND SIZE DIMENSIONS OF CONSTRUCTED COMPONENTS____ CONSTRUCTION NOTES____ DETAIL KEYS____ DRAWING CROSS-REFERENCES____ SPECIFICATION REFERENCES____.

DEMOLITION AND REPAIR

☐ EXISTING FENCES AND WALLS: WHICH TO RETAIN____ WHICH TO REPAIR____ WHICH TO REMOVE____.

☐ EXISTING STRUCTURES: WHICH TO RETAIN____ WHICH TO REPAIR____ WHICH TO RELOCATE____ WHICH TO DEMOLISH____.

☐ EXISTING PAVING, WALKS, STEPS, AND CURBS: WHICH TO RETAIN____ WHICH TO REPAIR____ WHICH TO REMOVE____.

☐ EXISTING ON-SITE UTILITIES, SEWERS, AND DRAINS: WHICH TO RETAIN____ WHICH TO USE____ WHICH TO REMOVE____.

☐ EXISTING TREES, SHRUBS, AND UNDERGROWTH: WHICH TO RETAIN____ WHICH TO REMOVE____ WHICH TO RELOCATE AND STORE FOR TRANSPLANTING____.

☐ AREAS FOR CLEARING AND GRUBBING____ EXISTING TRASH TO BE REMOVED____ STUMPS TO BE REMOVED____.

☐ ROCK OUTCROPS TO RETAIN____ ROCK TO BE REMOVED____.

GRADING, CONSTRUCTION, AND DRAINAGE

☐ EXISTING AND NEW SITE CONTOURS____ EXISTING AND NEW FINISH GRADES____ ELEVATION POINTS____.

☐ NEW BENCHMARK AND/OR BOUNDARY MARKERS.

☐ EXISTING HOLES AND TRENCHES: WHICH TO RETAIN____ WHICH TO FILL IN____.

☐ NEW FILL____ NOTE ON SOIL COMPACTION____.

☐ CUT AND FILL PROFILE (may be on separate drawing).

☐ NEW BUILDING AND RELATED STRUCTURES____ OVERALL EXTERIOR WALL DIMENSIONS____ DIMENSIONS TO PROPERTY

LINES____ OUTLINE OF FUTURE BUILDING ADDITIONS____.

☐ NEW BUILDING FINISH FLOOR ELEVATION AT GROUND FLOOR OR BASEMENT.

☐ NEW FINISH GRADE ELEVATIONS AT BUILDING CORNERS____ GRADE SLOPE AT BUILDING LINE____.

☐ FOUNDATION OR BASEMENT EXCAVATION LIMIT LINES.

☐ RETAINING WALLS OR WELLS FOR EXISTING TREES AFFECTED BY CHANGES IN FINISH GRADE: MATERIAL____ DIMENSIONS____ ELEVATION POINTS____ NOTE____ DETAIL KEYS____.

☐ NEW FENCES, GATES, AND WALLS: MATERIALS____ HEIGHTS____ DIMENSIONS____ DETAIL KEYS____.

☐ NEW RETAINING WALLS: MATERIAL____ HEIGHTS____ DIMENSIONS____ FOOTING LINES____ DRAINS____ DETAIL KEYS____.

☐ NEW SURFACE AND SUBSURFACE SOIL DRAINAGE: MATERIALS____ SIZES____ SLOPES____ DEPTHS____.

☐ NEW RIPRAP____ CULVERTS____ STORM DRAINS____ HEAD WALLS____.

☐ DRY WELLS: SIZES____ DEPTHS____.

☐ BUILDING PERIMETER FOUNDATION DRAIN TILE____ SLOPE AND DIRECTION OF DRAIN TO DRY WELL OR STORM SEWER____.

PAVING, WALKWAYS, AND PARKING

☐ NEW PUBLIC CURBS AND STREET DRIVEWAY ENTRIES.

☐ DRIVEWAYS: MATERIALS____ DIMENSIONS____ CENTERLINE ELEVATION POINTS____ DRAINAGE SLOPES____ DRAINS____.

☐ EXISTING AND NEW GRADE ELEVATIONS AT PAVING, NOTED AT CENTERLINES AND SIDES OF DRIVEWAYS.

☐ DRIVEWAY AND PARKING AREA GRADES (minimum 0.5%, maximum 5% at crowns).

☐ EXTERIOR PARKING LOCATED MINIMUM 5' FROM WALLS AND OTHER STRUCTURES____ PARKING LOCATED OUTSIDE TREE DRIP LINES____ PARKING LOCATED CLEAR OF ICICLE OR ROOF SNOW DROP LINES____.

☐ PAVEMENT AND WALKWAY ICE-MELTING EQUIPMENT.

☐ PAVEMENT HUMPS____ DIPS____ WARNING SIGNS TO SLOW THROUGH TRAFFIC____.

☐ CROSS-SECTIONAL DIAGRAMS THROUGH ROADWAYS OR DRIVEWAYS.

☐ RESTRICTED-ACCESS ROADWAYS____ LOCKABLE POSTS OR CHAIN BARRIERS AT RESTRICTED ACCESS____.

☐ PAINTED PARKING LINES____ TRAFFIC CONTROL LINES OR MARKERS____ ARROWS AND SIGNS PAINTED ON PAVEMENT____.

☐ PARKING BUMPERS____ TRAFFIC CONTROL CURBS____.

☐ VEHICULAR CURBS, BUMPERS, OR GUARD RAILINGS AT VULNERABLE AREAS: WALLS____ LEDGES____ WALKWAYS____ TREES____ STANDARDS____ COLUMNS____.

☐ WIDE-SPACE PARKING FOR HANDICAPPED____ RAMPS FROM PARKING AREA TO ADJACENT SIDEWALKS____.

☐ CONSTRUCTION JOINTS IN CONCRETE PAVING. (A typical arrangement is ½" joints every 20' to 30', plus joints at connections with other construction.)

☐ NEW WALKS AND STEPS: MATERIALS____ DIMENSIONS____ NOTES____ ELEVATION POINTS____ DIRECTION OF SLOPES____ RISER NUMBER AND HEIGHTS____.

☐ HANDRAILS AT STEPS OVER THREE RISERS.

☐ THREE RISERS MINIMUM AT ANY POINT ALONG WALKS OR RAMPS.

☐ ROUGH, NONSKID SURFACES AT EXTERIOR WALKS, STEPS, AND LANDINGS____ SLOPING WALK OR RAMP HANDRAILS____. (Maximum slope for walkway without handrails is 1 in 8.)

☐ PIPE SLEEVES AT PAVEMENT FENCING AND HANDRAIL POSTS.

☐ WALKWAY LOW POINTS SLOPED TO DRAIN.

☐ EXTERIOR STAIR TREADS SLOPED TO DRAIN____ SIDE GUTTERS AT RAMPS, STAIRS, OR STEPS____ CATCH BASINS OR DRAINS AT BASE OF STAIRS OR RAMPS____ IDENTIFICATION NUMBERS AT EXTERIOR STAIRS____ METAL TREAD NOSINGS AT CONCRETE STEPS____.

☐ CONCRETE WALKWAY CONSTRUCTION JOINTS. (Place ½" joints each 30', tooled joints at 5'. Construction joints at connections with other construction.)

☐ PAVED TERRACES____ PATIOS____ OFF-THE-GROUND DECKS____. (Open joint drainage or closed joint pavers. Solid paving sloped minimum 1" in 10' for drainage.)

☐ OPENINGS OR GRATINGS AT TREES IN PAVED AREAS FOR WATERING SPACE.

☐ SYNTHETIC GRASS OR TURF.

☐ SIDEWALK VAULTS____ SIDEWALK PAVING GLASS FOR UNDERGROUND ROOMS____ SIDEWALK ELEVATOR____ FUEL DELIVERY ACCESS____.

- ☐ SIDEWALK PLAQUES SPECIFYING PROPERTY OWNERSHIP AND LIMITS ON PUBLIC USE.
- ☐ MANHOLES____ GRATINGS____ CLEAN-OUTS____ METER BOXES____.
- ☐ BICYCLE AND PARAPLEGIC RAMPS ADJACENT TO WALKWAY STEPS AND AT CURBS.

LANDSCAPING

- ☐ EXISTING TREES, SHRUBS, AND UNDER-GROWTH: WHICH TO RETAIN____ WHICH TO REMOVE____ WHICH TO RELOCATE AND STORE FOR TRANSPLANTING____
- ☐ TREES TO BE WRAPPED____ TREE GUARDS FOR PROTECTION DURING CONSTRUCTION____.
- ☐ NEW TOPSOIL STORAGE AREA____ EXCAVATION LOAM STORAGE AREA____.
- ☐ TEMPORARY EROSION CONTROL.
- ☐ NEW LANDSCAPE AREAS WITH LANDSCAPE LEGEND AND KEYS TO IDENTIFY TREE, PLANT, AND LAWN TYPES.
- ☐ SLOPES AND DRAINAGE FOR LAWN AND GROUND COVER AREAS____ ALL POCKETS AND STANDING WATER AREAS FILLED OR PROVIDED WITH DRAINAGE____.
- ☐ PLANTING BEDS____ BARK____ GRAVEL____ CURBS OR BORDER BOARDS AT BEDS____.
- ☐ ROOT FEEDING PIPES.
- ☐ LANDSCAPE IRRIGATION AND/OR SPRINKLER SYSTEM: AUTOTIMER____ VALVES____ HOSE BIBBS AND HYDRANTS____ SPRINKLER HEAD LOCATIONS____. (May be shown on separate plumbing plan.)
- ☐ DECORATIVE YARD LIGHTING AND TIMER SWITCH. (May be shown on electrical drawings.)

APPURTENANCES AND SITE FURNITURE

This is often combined with the landscaping plan.

- ☐ NEW PONDS, POOLS, FOUNTAINS: MATERIALS____ UTILITIES____. (Fountains located away from wind channels.)
- ☐ NEW PLANTERS____ PLANT TUBS____ PLANTER DRAINS____.
- ☐ SCULPTURE PEDESTALS OR SUPPORT SLABS.
- ☐ DIRECTION SIGNS____ LOCATION MAP STANDS____ DIRECTORY PLAQUES____ BULLETIN BOARDS____ KIOSKS____.
- ☐ WALKWAY AND PARKING LIGHT STANDARDS AND PEDESTALS.
- ☐ BENCHES (usually set 2′ back from walkways).
- ☐ TRASH RECEPTACLES____ PUBLIC ASH-TRAYS____.

- ☐ PERGOLAS____ LATTICES____.
- ☐ DRINKING FOUNTAINS____ PARAPLEGIC FOUNTAINS____ CHILDREN'S FOUNTAINS____ PET FOUNTAINS____.
- ☐ BICYCLE RACKS____ PRAM RACKS____.
- ☐ NEWSPAPER RACKS.
- ☐ PUBLIC PHONES____ EMERGENCY PHONES____ LOW PHONE FOR PARAPLEGICS____ FIRE ALARM BOXES____.
- ☐ PUBLIC MAILBOX____ CLUSTER MAILBOXES____ MAIL SLOT FOR PARAPLEGICS____.
- ☐ FLAGPOLES.
- ☐ BUS OR SHUTTLE STOP: BENCH____ SHELTER____ WASTE RECEPTACLE(S)____.
- ☐ WALKWAY CANOPIES____ WIND SCREENS____.
- ☐ TURNSTILES.
- ☐ INFORMATION OR GUARD BOOTH.
- ☐ PAID PARKING: TICKET DISPENSER____ AUTOMATIC GATE____ TIRE SPIKE BARRIER____.
- ☐ MAINTENANCE YARD____ GASOLINE STORAGE TANK____ GASOLINE TANK INLET, VALVE, AND PUMP____.
- ☐ TRASH ENCLOSURES____ INCINERATOR YARD ENCLOSURE____.
- ☐ PLAYGROUND EQUIPMENT____ FENCED PLAY YARD____.

UTILITIES

- ☐ NEW WATER SUPPLY MAIN____ NEW WATER METER____ WATER METER AND SHUTOFF BOX AND COVER____.
- ☐ FIRE HYDRANTS____ HOSE BIBBS____ STANDPIPES____.
- ☐ NEW SEWER MAIN____ VENT AND CLEAN-OUT____ DIRECTION OF SLOPE____ ELEVATION POINTS____.
- ☐ NEW GAS MAIN____ GAS METER____ SHUTOFF VALVE AT BUILDING (conspicuous location for shutoff, with identifying sign)____.
- ☐ NEW BURIED OR OVERHEAD ELECTRIC CABLE____ PHONE CABLE____ TV CABLE____.
- ☐ NEW POWER POLES. (Overhead cables to clear walks and driveways as required by building authority.)
- ☐ ALL UTILITY TRENCHES: MINIMUM AND MAXIMUM DEPTH LIMITS____ TRENCHING LOCATED TO AVOID DAMAGE TO TREE ROOTS AND NEIGHBORING STRUCTURES____.
- ☐ EXCAVATION WARNING SIGNS/PLAQUES ALONG ROUTES OF BURIED UTILITY LINES.
- ☐ LENGTHS OF MAINS.
- ☐ EXTERIOR ELECTRIC OUTLETS AND LIGHTS AT

BUILDING AND AT AUXILIARY STRUCTURES.

☐ FUEL OIL OR LIQUIFIED GAS STORAGE TANK____ FUEL LINE TO BUILDING____ VENT____ GAUGE BOX____ FILL BOX____ MANHOLE____.

☐ PACKAGE SEWAGE TREATMENT TANK AND HOUSING: SIZE____ LOCATION____ CONSTRUCTION NOTES____ DETAIL KEYS____.

CONSTRUCTION WORK SITE PLAN

☐ EROSION CONTROL____ TEMPORARY RETAINING WALLS____ SHORING____ TEMPORARY DRAINAGE____.

☐ JOB SIGN AND TEMPORARY FENCE: DESIGN AND DETAILS.

☐ CONSTRUCTION SHED____ BARRIERS____ GATES____ TOILETS____.

☐ TESTING AREA____ SAMPLES STORAGE AREA____.

☐ MATERIAL AND EQUIPMENT STORAGE AREAS____ PLATFORMS____.

☐ TEMPORARY ROADS____ WALKWAYS____.

☐ TOPSOIL STOCKPILE AREA.

☐ PLANT STORAGE YARD____ GREENHOUSE____.

☐ CONSTRUCTION POWER POLE AND METER.

☐ SITE ILLUMINATION.

☐ SECURITY GUARD'S SHANTY.

☐ CONSTRUCTION WATER SUPPLY HYDRANTS____ TEMPORARY HOSE BIBBS____.

☐ BRIDGES FOR PROTECTION OF PAVING, WALKS, CURBS FROM CONSTRUCTION EQUIPMENT AND TRUCKS.

☐ TRASH AND DEBRIS STORAGE.

☐ SOIL TEST-BORING PLAN____ TEST PITS____.

FLOOR PLANS

GENERAL REFERENCE DATA

☐ DRAWING TITLES AND SCALE.

☐ ARROWS SHOWING COMPASS NORTH AND REFERENCE NORTH.

☐ MODULAR GRID OR STRUCTURAL COLUMN GRID WITH NUMBER AND LETTER COORDINATES.

☐ SQUARE FOOTAGE TOTALS: BUILDING____ AUXILIARY STRUCTURES____ DECKS AND BALCONIES____.

☐ ROOM NAMES OR NUMBERS____ EXTERIOR AREAS____.

☐ INTERIOR ELEVATION ARROW SYMBOLS AND REFERENCE NUMBERS.

☐ EXTERIOR ELEVATION ARROW SYMBOLS AND REFERENCE NUMBERS.

☐ OVERALL CROSS-SECTIONAL LINES AND KEYS.

☐ MATCH-UP LINE, OVERLAP LINE, AND REFERENCE IF FLOOR PLAN IS CONTINUED ON ANOTHER SHEET.

☐ SPACE FOR ITEMS N.I.C.____ NOTE ON N.I.C. ITEMS TO BE INSTALLED, CONNECTED BY CONTRACTOR____.

☐ REMODELING: EXISTING WORK TO BE REMOVED____ EXISTING WORK TO REMAIN____ EXISTING WORK TO BE REPAIRED OR ALTERED____.

☐ GENERAL NOTES____ BUILDING CODE REFERENCES____.

☐ MATERIALS HATCHING OR POCHÉ.

☐ CONSTRUCTION NOTES____ DETAIL KEYS____ DRAWING CROSS-REFERENCES____ SPECIFICATION REFERENCES____.

WALLS, PARTITIONS, AND COLUMNS

☐ EXTERIOR WALLS____ COLUMNS AND POSTS____ WALL OPENINGS____.

☐ OUTLINE OF FUTURE BUILDING ADDITIONS.

☐ INTERIOR WALLS____ PARTITIONS____ COLUMNS AND POSTS____.

☐ MOVABLE WALLS____ DEMOUNTABLE PARTITIONS____ PORTABLE CUBICLES____.

☐ WIRE MESH PARTITIONS____ WOOD OR METAL GRILLE PARTITIONS____.

☐ GLAZED WALLS: OBSCURE GLASS___ WIRE GLASS___ PLASTIC___ ACOUSTICAL GLASS___ SURVEILLANCE MIRRORS___. (For other components, see Windows.)

☐ OUTLINE AND NOTED HEIGHTS OF LOW PARTITIONS___ SCREENS___ PREFAB STORAGE WALLS___ DIVIDERS___ PLANTER WALLS___.

☐ DOOR AND WINDOW OPENINGS (see door and window sections).

☐ OUTLINE OF CONSTRUCTED VENEERS___ MATERIALS___ HEIGHT OF VENEERS___ PANELING___ DADOS___ WAINSCOTS___ TILE WORK___.

☐ BROKEN-LINE OUTLINES: OVERHANGS___ CANOPIES___ OVERHEAD BALCONIES___ COVES___ VALANCES___ OTHER PROJECTIONS___. (May be on reflected ceiling plan.)

☐ DASHED WALL LINES AT FRAMED OPENINGS.

☐ WALL OUTLINES: NOTE WHETHER FINISH SURFACES, ROUGH SURFACES, OR LINES OF FRAMING ARE REPRESENTED.

☐ VERTICAL IN-WALL CHUTES___ FURRED WALLS___ PNEUMATIC TUBES___ WIRING, VENT, AND PLUMBING CHASES___.

☐ INTERIOR RAINWATER LEADERS FROM ROOF.

☐ INDICATION FOR WALLS THAT PASS THROUGH SUSPENDED CEILINGS___ WALLS THAT STOP AT SUSPENDED CEILINGS___.

☐ WALL SCUTTLES___ ACCESS PANELS___ GRILLES___ SLEEVES AND OTHER THRU-WALL OPENINGS___.

☐ NOISE BARRIER CONSTRUCTION WHERE CONVECTORS PASS THROUGH PARTITIONS.

☐ SOUND-ISOLATION WALLS___ VIBRATION AND NOISE ISOLATION AT WALL-MOUNTED TVs, INTERCOMS, FANS, AND OTHER MOVING OR NOISE-MAKING EQUIPMENT___.

☐ NOTES AND INDICATION OF WALLS WITH MEMBRANE WATERPROOFING.

☐ RADIO SHIELDING___ RADIATION SHIELDING___ EXPLOSION PROOFING___.

☐ ACOUSTICAL PLASTER___ BATT INSULATION___ RIGID INSULATION___ POURED INSULATION___.

☐ NOTES AND INDICATION OF FIRE-RESISTIVE CONSTRUCTION AT WALLS AND COLUMNS.

☐ PROTECTIVE RAILS, BUMPERS, OR CORNER GUARDS AT WALLS, GLAZING, COLUMNS, OR DOOR JAMBS SUBJECT TO DAMAGE.

☐ PROTECTIVE JACKETS WHERE FIRE-RESISTIVE CONSTRUCTION IS SUBJECT TO DAMAGE.

☐ WATER PROTECTION AT WALLS BEHIND DRINKING FOUNTAINS AND WORK SINKS.

☐ WALL-MOUNTED SHELVES___ CASEWORK___ MIRRORS___ PEGBOARDS___ TACKBOARDS___ CHALKBOARDS___ MAP RAILS___ PROJECTION SCREENS___ HOOKS___ HANGERS___ RACKS___ SIGNS___.

☐ WALL-MOUNTED CLOCKS___ LIGHTS___ TVs___ INTERCOMS___ TIME CLOCKS___ ALARMS___ SPEAKERS___ THERMOSTATS___ FIRE EXTINGUISHERS___ INTRUDER DETECTORS___ SMOKE AND FIRE DETECTORS___.

☐ WALL-MOUNTED COUNTERS___ WORKBENCHES___.

☐ WALL SAFES___ VAULT WALLS___.

☐ WALL-MOUNTED MURALS___ BAS-RELIEF___ FRIEZES___ SCULPTURE___.

☐ CORNERSTONE___ ARCHITECT'S PLAQUE___.

DIMENSIONING

☐ NOTES ON DIMENSIONING SYSTEM: MODULAR DIMENSIONING PRACTICE AND CONVENTIONS___ WHETHER DIMENSIONS ARE TO FINISH SURFACES, TO ROUGH SURFACES, OR TO ROUGH FRAMING___ NOTES OF SPECIAL ARROWHEAD INDICATIONS IF DIMENSIONING SYSTEM IS MIXED___.

☐ DIMENSIONS NOTED: N.T.S.___ "VARIES"___ ±___ "EQ."___ "TYP."___.

☐ FINISH GRADE ELEVATION POINTS AT CORNERS AND PERIMETER OF BUILDING.

☐ OVERALL EXTERIOR WALL LENGTHS___ OFFSETS___ PROJECTIONS___ RECESSES___.

☐ SIZES OF REMOVABLE EXTERIOR WALL PANELS.

☐ EXTERIOR OPENINGS. (Openings that clearly center on bays or are adjacent to columns often don't require dimensioning; door and window sizes may be covered in schedules.)

☐ INTERIOR WALL AND PARTITION LENGTHS___ DISTANCES BETWEEN INTERIOR WALLS OR PARTITIONS IN DIMENSION STRINGS ACROSS LENGTH AND WIDTH OF BUILDING___. (Dimension strings are located so as not to conflict with other essential notes, such as room names or numbers. Parallel strings of dimensions on interiors aren't considered desirable.)

☐ COLUMN, POST, MULLION CENTERLINE DIMENSIONS___ CUMULATIVE DIMENSION NOTES AT COLUMN LINES___.

☐ DIMENSIONS OF OVERHANGS AND OVERHEAD PROJECTIONS.

☐ THICKNESSES OF EXTERIOR WALLS AND INTERIOR PARTITIONS. (Interior partitions are often dimensioned to centerlines.)

217

☐ DIMENSIONS OR DIMENSION NOTES: CHASES___ FURRED WALLS___ SHAFTS___ FUTURE SHAFTS___.

☐ SMALL THRU-WALL OPENINGS___ LENGTH AND HEIGHT DIMENSIONS___ HEIGHT OF SILLS OF OPENINGS ABOVE FLOOR___.

☐ BUILT-INS___ SIZES___ DIMENSIONS TO FINISHED SURFACES___.

☐ HEIGHTS: RAILS___ COUNTER TOPS___ BUILT-INS___ FIXTURES___.

☐ SPACINGS OF PLUMBING FIXTURES.

☐ RADII OF ARCS WITH REFERENCE DIMENSIONS TO LOCATE CENTERS OF ARCS OR CIRCLES.

☐ ANGLES IN DEGREES OF NON-RIGHT-ANGLE WALLS AND BUILT-INS. (In lieu of degree measurements, ends of angled unit or wall may be dimensioned from reference line or wall.)

☐ CENTERLINES OF SYMMETRICAL PLANS WITH NOTE THAT DIMENSIONS ARE THE SAME AT BOTH SIDES OF CENTERLINE.

WINDOWS

☐ GLASS LINES___ SILL AND STOOL LINES___ MULLIONS___.

☐ INTEGRATED STOREFRONT SYSTEM___ METAL-FRAME SCHEDULE___ DETAIL KEYS___.

☐ OPERABLE DIRECTION OF HORIZONTAL SLIDING WINDOWS___ SWINGS FOR VERTICAL CASEMENTS___ PIVOT WINDOWS___ AWNING WINDOWS___.

☐ WINDOW SYMBOLS KEYED TO WINDOW SCHEDULE.

☐ FIXED GLASS___ OBSCURE GLASS___ WIRE GLASS___ PLASTIC___ SAFETY GLASS___ ACOUSTICAL GLASS___ INSULATED GLASS___ STAINED GLASS___ TEMPERED GLASS___ DOUBLE GLAZING___. (May be noted only in window schedule.)

☐ EXTERIOR SHUTTERS___ GRILLE WORK___ SUN-CONTROL DEVICES___.

☐ PROTECTIVE GRILLES AT WINDOWS ACCESSIBLE TO PUBLIC WALKWAYS, EXTERIOR STAIRS, OR FIRE ESCAPES___ FIXED OR OPERABLE WIRE MESH WINDOW GUARDS___.

☐ WINDOW INTRUDER ALARMS___ SECURITY WINDOWS___.

☐ WINDOW CLEANER'S HOOKS.

☐ WINDOW DETAIL REFERENCE KEYS (if not covered at window elevation drawings with schedule).

☐ INTERIOR DRAPE OR WINDOW SHADE TRACKS___ FOLDING SHUTTERS OR SCREENS___ VENETIAN BLIND POCKETS___.

☐ WINDOW OPENINGS DESIGNED AND LOCATED TO AVOID CONFLICT OF MOLDING, TRIM, SILLS, OR STOOLS WITH ADJACENT WALLS, WINDOWS, OR DOORS.

☐ COUNTER WINDOWS___ GLASS PASS-THRUS___ INTERIOR GLAZED PARTITIONS (fixed glass provided with removable stops on one side)___.

☐ CAULKING AT ALL SASH IN MASONRY WALLS.

☐ SPANDREL GLASS INTEGRATED WITH FENESTRATION.

DOORS

☐ DOOR OPENINGS___ DOOR SWING DIRECTIONS___ SYMBOL KEY REFERENCES TO DOOR SCHEDULE___ SOLID-CORE EXTERIOR DOORS___ HOLLOW-CORE INTERIOR DOORS___ DOOR THICKNESSES___ VENEER NOTES___.

☐ DOOR SYMBOL SYSTEM FOR ELEVATOR DOORS.

☐ SLIDING AND HINGED SCREEN DOORS (usually shown in broken lines).

☐ LARGE DOORS: HEADER OR LINTEL MATERIALS AND SIZES (usually noted in separate structural drawings).

☐ REMOVABLE WINTER AIR LOCK STORM DOORS AT ENTRY VESTIBULES.

☐ REMOVABLE AND LOCKABLE STILES FOR EXTERIOR DOUBLE DOORS.

☐ STRIKE SIDE OF DOORS: LOCATED GENEROUS DISTANCE FROM COLUMNS, WING WALLS, WINDOW MULLIONS, OTHER VULNERABLE CONSTRUCTION.

☐ DOOR JAMBS: DESIGNED AND LOCATED TO AVOID CONFLICT OF MOLDINGS OR TRIM THAT INTERSECT ADJACENT WALLS, DOORS, OR WINDOWS.

☐ PROTECTIVE METAL JACKETS FOR DOOR OPENINGS SUBJECT TO DAMAGE.

☐ DOOR OPENINGS: LOCATED SO THAT DOORS AVOID TANGLING WITH NEIGHBORING DOORS IN ADJACENT WALLS.

☐ DOORS OPENING ON OPPOSITE SIDES OF COMMON CORRIDOR: STAGGERED IN LOCATION FOR VISUAL AND SOUND PRIVACY.

☐ DOORS OPENING TO TOILET ROOMS, DRESSING ROOMS, OTHER PRIVATE AREAS: LOCATED AND HINGED TO BLOCK DIRECT VIEW INTO ROOMS.

☐ SADDLE, JAMB, OR BUCK DETAIL KEYS OR NOTES IF NOT SHOWN ON SCHEDULE. (Metal bucks sometimes shown at door jambs on plan.)

☐ SADDLES OR SILLS___. WEATHERSTRIPPING

AT EXTERIOR DOORS____ SADDLE MATERI-AL____.

☐ SOUNDPROOFING AND STRIPPING AT THRESHOLDS OF NOISY ROOMS.

☐ WEATHERSTRIPPING AT PLENUM SPACE ACCESS DOORS.

☐ PANIC BARS____ DOOR UNDERCUTS____ LOUVERS____ KICKPLATES____ "PUSH-PULL" PLATES____ VIEW PANELS____ CLOSERS____ MIRRORS____. (These may be covered in specifications or schedules.)

☐ ALARMS AND WARNING SIGNS AT RESTRICTED EXIT DOORS.

☐ DOOR-MOUNTED SIGNS____ NAME AND NUMBER PLATES____.

☐ FIXED MATCHING PANELS ABOVE WOOD FLUSH DOORS.

☐ FLOOR CUTS FOR PIVOT-HINGE DOORS____ FLOOR-MOUNTED DOORSTOPS____.

☐ DOOR ROLLER BUMPERS WHERE ADJACENT DOORS MAY HIT EACH OTHER.

☐ FLOOR-MOUNTED TRACKS FOR SLIDING DOORS____ SLIDING PANELS____ BI-FOLD PANELS____.

☐ OVERHEAD TRACKS FOR FOLDING AND SLIDING DOORS.

☐ SURFACE-MOUNTED DOORS.

☐ DOUBLE-ACTING DOORS____ KALAMEIN DOORS____ METAL-CLAD DOORS____ WOVEN WIRE MESH DOORS____ DUTCH DOORS____ GATES____ FLEXIBLE DOORS____ INVISIBLE DOORS IN PANELING____.

☐ AUTOMATIC DOORS AND DOOR-ACTIVATING EQUIPMENT.

☐ DIRECTION OF MOVEMENT: SLIDING____ FOLDING____ BI-FOLD____ PIVOTED____ DOUBLE-ACTING____ VERTICAL____.

☐ SLIDING DOOR POCKETS (located to avoid conflicts with structure or electrical work).

☐ OVERHEAD FOLD-DOWN DOORS____ SECTIONAL DOORS____ ROLLING METAL DOORS____ SIDE-COIL DOORS____ ROLLING GRILLES____.

☐ AIR DOORS.

☐ FUSIBLE-LINK HINGED OR SLIDING FIRE DOORS.

☐ SIDEWALK DOORS.

FLOORS: GENERAL

☐ FINISH FLOOR MATERIAL INDICATION AND/OR NAME (sometimes noted by type under each room name or number and described in detail in finish schedule and/or specifications).

☐ DIVIDING LINES AND FLOOR HEIGHT NOTES AT CHANGES IN FLOOR LEVEL. (Note whether floor elevations apply to slab, subfloor, or finish floor.)

☐ STEPS: DIRECTION ARROW____ NUMBER OF RISERS____ HANDRAIL____.

☐ RECESSED SLAB FOR TILE, MASONRY, OR TERRAZZO____ HEIGHTS SET FOR ALIGNMENT OF DIFFERENT FINISH FLOOR SURFACES____.

☐ JOINT PATTERNS OF MASONRY, TILE, MARBLE, OR TERRAZZO FLOORS. (Smooth surfaces should be avoided at landings, entries, and vestibules.)

☐ ROUGH SURFACES NOTED AT ENTRY LANDINGS, EXTERIOR STEPS.

☐ EXPANSION JOINTS, TOOLED JOINTS, CONSTRUCTION JOINTS, AND PERIMETER JOINTS IN SLABS____ JOINT DETAIL KEYS____.

☐ EXPANSION SPACE AT PERIMETER OF WOOD STRIP OR WOOD BLOCK FINISH FLOORING.

☐ SADDLES, THRESHOLDS, METAL STRIPS SEPARATING ONE FLOOR MATERIAL FROM ANOTHER.

☐ UNDERLAYMENT FOR RESILIENT FLOORS.

☐ BUILT-IN EQUIPMENT AND BUILT-IN FLOOR-MOUNTED FURNISHINGS: FLOOR ANCHORS FOR BUILT-INS AND EQUIPMENT____ ANCHORS FOR N.I.C. EQUIPMENT____ FUTURE EQUIPMENT____.

☐ BALCONY, EXTERIOR LANDING, DECK: FLOOR MATERIAL____ FLOOR ELEVATION POINTS____ DIRECTION OF SLOPE____ LANDING AND BALCONY FLOOR DRAINS____ EXTERIOR SURFACE ELEVATION 3″ BELOW DOOR THRESHOLDS____.

☐ FLOOR OPENINGS FOR ACCESS PANELS____ HINGED OR REMOVABLE PANELS____.

☐ WATERPROOF MEMBRANE CONSTRUCTION AT SHOWER AND OTHER WET ROOMS.

☐ RAISED FLOORS____ PREFABRICATED PLATFORM FLOORS____.

☐ CABLE, PIPE, AND DRAIN TRENCHES____ METAL FRAMES AND COVERS____.

☐ THRU-FLOOR SHAFTS____ CHASES____ CHUTES____ FUTURE SHAFTS OR CHASES____ SHAFT NUMBERS____.

☐ CAPPED SLEEVES FOR MOVABLE STANCHIONS, OTHER PORTABLE FLOOR-MOUNTED FIXTURES OR EQUIPMENT.

☐ SOUND- AND VIBRATION-ISOLATION PADS FOR POWERED EQUIPMENT____ ISOLATION SLABS____.

☐ RECESSED DOORMATS____ METAL FRAME AND DEPTH OF MAT SINKAGE____.

☐ FLOOR DRAINS AT MECHANICAL ROOMS, GARAGES, TRASH ROOMS. (A generous number of floor drains is recommended.)

- ☐ DATUM REFERENCES AT INTERIOR BALCONIES, LANDINGS, AND MEZZANINES.
- ☐ CONDUCTIVE FLOORS.
- ☐ WOOD DECKING OVER FLOOR.
- ☐ METAL ARMORING GRIDS.
- ☐ CATWALKS____ 60° SHIP'S LADDERS____ PIT AND MANHOLE STEP RUNGS____ SPIRAL LADDERS AND STAIRS THROUGH FLOORS____ CAGED LADDERS____.
- ☐ THRU-FLOOR CONVEYORS____ DUMBWAITERS____ BOOK LIFTS____ MONEY LIFTS____ PEOPLE LIFTS____ COAL CHUTE SLIDE GUIDES____ TURNTABLES____ ELEVATOR AND CONVEYOR SHAFT NUMBERS____.

FLOORS: CONCRETE SLAB ON GRADE

Also see Floors: General.

- ☐ SLAB PERIMETER OUTLINE AND DIMENSIONS.
- ☐ PERIMETER AND INTERIOR FOOTINGS SHOWN IN BROKEN LINES.
- ☐ FOOTINGS: PIERS____ COLUMNS____ PILASTERS____ STAIRS____ BEARING WALLS____.
- ☐ FOOTING SIZE AND LOCATION DIMENSIONS.
- ☐ ELEVATION POINTS AT BOTTOMS OF FOOTINGS.
- ☐ EXISTING AND NEW FINISH GRADE LINES____ FINISH GRADE ELEVATION POINTS AT PERIMETER AND CORNERS OF BUILDING____.
- ☐ DRAIN TILE WITH CRUSHED-ROCK BED.
- ☐ PERIMETER THERMAL INSULATION AT JUNCTURE OF SEPARATELY POURED SLAB AND FOOTING.
- ☐ TERMITE SHIELDS____ RODENT SHIELDS____.
- ☐ ADJACENT SLABS____ SIZE AND SPACING OF DOWEL CONNECTIONS BETWEEN SLABS____.
- ☐ FILLED, PUDDLED, AND TAMPED SOIL UNDER RAISED CONCRETE LANDINGS.
- ☐ CONCRETE PATIO OR TERRACE SLAB FROST CURB (6″ edge of slab turned down 12″ or more below finish grade).
- ☐ CONCRETE SLAB CONTROL JOINTS, CONSTRUCTION JOINTS: ½″ CONSTRUCTION JOINTS EVERY 40′ BOTH WAYS____ TOOLED JOINTS EVERY 20′ BOTH WAYS AND AT PIERS AND COLUMNS____ SAWED RELIEF JOINTS ONE-FIFTH DEPTH OF SLAB____.
- ☐ BROKEN-LINE OUTLINE OF FLOOR PLAN ON SLAB, INCLUDING PLUMBING FIXTURES.
- ☐ CHANGES IN SLAB FLOOR ELEVATION.

- ☐ SLAB DEPRESSIONS FOR MASONRY, TILE, OR TERRAZZO FINISH FLOORING.
- ☐ UNDER-FLOOR OR PERIMETER DUCTWORK.
- ☐ RADIANT-HEATING PIPE PLAN. (May be separate drawing.)
- ☐ CANTILEVERED SLAB SECTIONS____ CHANGE IN TENSILE REINFORCING FROM BOTTOM TO UPPER PORTION OF SLAB NOTED____.
- ☐ SLAB THICKNESS____ REINFORCING____ VAPOR BARRIER____ ROCK SUB-BASE____ SUBGRADE COMPACTION____.
- ☐ PARTIAL PLAN OF FINISH FLOOR: MATERIAL____ SLEEPERS____ PATTERN____ SLOPE____ THICKNESS OR HEIGHT OF FLOORING ABOVE SLAB____.
- ☐ DETAIL SECTION KEYS: FOOTINGS____ CURBS____ CONNECTIONS TO OTHER CONSTRUCTION____ POST AND COLUMN CONNECTIONS AT SLAB OR PIERS____.

ROOMS AND SPACES: FLOOR PLANS AND INTERIOR ELEVATIONS

ENTRY LOBBY AND CORRIDORS

- ☐ BUILDING NAME BESIDE OR OVER ENTRY____ BUILDING ADDRESS NUMBER____ ENTRANCE SIGNS____.
- ☐ MAIN ENTRY SPOTLIGHTING____ PAVEMENT UP-LIGHTS____.
- ☐ NONSLIP ENTRY LANDINGS____ ROOF OVERHANGS OR CANOPY WEATHER PROTECTION____.
- ☐ DRAWING ARROWS OR NOTES INDICATING MAIN ENTRY AND SECONDARY ENTRANCES____ VESTIBULE NUMBERS____.
- ☐ WEATHER- AND SKID-RESISTIVE VESTIBULE FLOOR SURFACES.
- ☐ MAT SINKAGE AT ENTRY DOORS.
- ☐ AUTOMATIC DOOR OPENERS____ FLOOR RECESS FOR DOOR-ACTIVATING EQUIPMENT____ DOOR SWING GUARDRAILS____.
- ☐ GUARDRAILS AT FULL-HEIGHT GLASS WALLS. (Human collisions with glass are most common near entrances.)
- ☐ PUSH-PULL PLATES ON PUBLIC DOORS____ HEAVY-DUTY CLADDING AT DOORS USED FOR DELIVERIES____.
- ☐ NIGHT DEPOSITORY SLOT AND BOX.
- ☐ PUBLIC ASHTRAYS____ TRASH RECEPTACLES____. (Near entrances and at elevator lobby.)
- ☐ BENCHES____ BUILT-IN WRITING OR READING TABLES____ DISPLAY TABLES____ BROCHURE

RACKS____.

☐ WALL-HUNG DISPLAY CABINETS.

☐ PICTURE RAIL AND ADJUSTABLE LIGHTING FOR DISPLAYS.

☐ BUILT-IN TREE OR SHRUB POTS____ RECESSED PLANTERS____ DRAINS____ WATER SUPPLY____.

☐ DIRECTORY____ DIRECTION SIGNS____ BULLETIN BOARDS____.

☐ DRINKING FOUNTAINS____ AUXILIARY FOUNTAIN FOR PARAPLEGICS____ WALL PROTECTION BEHIND FOUNTAINS____.

☐ MAIL DEPOSITORY____ MAIL CHUTES AT CORRIDORS OF MULTISTORY BUSINESS BUILDINGS____ POSTAGE STAMP MACHINE____ WRITING AND PACKAGE SHELVES____.

☐ VENDOR'S ALCOVE____ CLOSABLE COUNTER OR SNACK BAR____.

☐ TELEPHONE ALCOVES____ LOW PHONE SHELF FOR PARAPLEGICS____.

☐ VENDING MACHINE ALCOVES WITH WASTE RECEPTACLES.

☐ DRAPE OR BLIND VALANCES____ AUTOMATIC CONTROLS____.

☐ CAPPED FLOOR SLEEVES FOR ROPE STANCHIONS____ WALL HOOKS FOR ROPE CONNECTIONS____.

☐ DELIVERY TRAFFIC AREAS: BUMPERS OR RAILS AT GLASS WALLS____ WALL CORNER GUARDS____ CURBS OR PROTECTIVE BASE AT WALLS____ LIMIT LINES ON SERVICE CORRIDOR FLOORS____.

☐ ELEVATOR LOBBY: NONSKID FLOOR____ DRINKING FOUNTAIN____ ASHTRAYS____ TRASH RECEPTACLES____ LOW-HEIGHT CALL BUTTONS FOR PARAPLEGICS____.

☐ FIRE ALARM BOXES____ EXTINGUISHER AND FIRE HOSE CABINETS____ ALARMS____.

☐ EMERGENCY CALL BUTTONS.

☐ CLOCKS____ SPEAKERS____ ANNUNCIATORS____.

☐ MOUNTS FOR CLOSED-CIRCUIT TV CAMERAS.

☐ SECURITY GUARD'S CLOCKS.

☐ ILLUMINATED EXIT SIGNS OVER DOORS.

☐ THERMOSTATS____ UNIT VENTILATORS____ UNIT HEATERS____ CONVECTORS AND ENCLOSURES____ REGISTERS____.

TOILET ROOMS

☐ DRINKING FOUNTAIN OUTSIDE RESTROOM ENTRIES.

☐ SMOOTH, NONABSORBENT FLOOR SURFACE, UP WALL TO 4' MINIMUM. (Terrazzo is subject to deterioration near urinals and water closets.)

☐ FLOOR SLOPES AND DRAINS.

☐ HOSE BIBB____ JANITOR'S SERVICE SINK____.

☐ LAVATORIES (30″ spacing typical)____ PARAPLEGIC LAV____ WASH FOUNTAIN____.

☐ URINAL STALLS (30″ spacing typical)____ WIDE STALL WITH GRAB BARS____ URINAL PARTITIONS (metal partitions are subject to corrosion)____.

☐ WATER CLOSETS (32″ spacing typical)____ WIDE PARAPLEGIC STALL WITH DOUBLE-SWING DOORS AND GRAB BARS____.

☐ WATER CLOSET STALL ACCESSORIES: RIGID OR FOLD-DOWN PACKAGE AND PURSE SHELF____ TOILET PAPER HOLDERS____ SANITARY NAPKIN DISPENSER____ NAPKIN DISPOSER____ SEAT COVER TISSUE DISPENSER____ COAT HOOK____ ASHTRAY____.

☐ PAPER TOWEL OR ROTARY TOWEL DISPENSERS____ WARM-AIR HAND DRYERS____.

☐ WASTE PAPER RECEPTACLES____.

☐ LAV SHELVES____ LAV MIRRORS____ ELECTRIC RAZOR OUTLET____.

☐ FULL-LENGTH MIRROR____ COAT AND UMBRELLA HOOKS____.

☐ VENDING MACHINE AND SCALE ALCOVE____.

☐ ACCESS PANELS AT WATER CLOSETS TO PLUMBING CHASE.

☐ COMBINATION WATER CLOSET STALL AND DRESSING ROOM WITH BENCH, CLOTHES HOOKS, AND MIRROR.

LAUNDRY ROOM

☐ OVERFLOW CURBS____ FLOOR DRAINS____.

☐ FREE-SWING, DOUBLE-ACTING DOORS____ VIEW PANELS AT DOORS____.

BOILER ROOM

☐ REMOVABLE WALL PANELS FOR CHANGE OF BOILER UNITS.

☐ CRANE HOIST.

☐ VENTILATED PLENUM ABOVE BOILER ROOMS AS REQUIRED TO INSULATE ROOMS ABOVE.

☐ BOILER FLOOR DROP OR PIT____ FLOOR PLATES____ FLOOR DRAINS____.

☐ GUARDRAILS AT PITS____ 60° SHIP'S LADDERS____ RAISED WALKWAYS____ CATWALKS____ CRANE HOIST____.

ELEVATOR AND DUMBWAITER PITS

☐ GUARDRAIL____ PIT LADDER____ FLOOR

221

DRAIN____.

☐ LIGHT SWITCH AND CONVENIENCE OUTLET. (Maintenance outlets also provided in elevator shaft.)

JANITOR'S CLOSET, MAINTENANCE SHOP

☐ SMOOTH, NONABSORBENT FLOOR SUR-FACE____ FLOOR SINK____.

☐ UTILITY SHELVES____ RACKS____ CLOTHES HOOK____.

☐ DOOR CLADDING____ DOOR JAMB CORNER GUARDS____.

☐ CLOTHES LOCKERS____ SHOWER STALLS____ BENCHES____.

☐ PIPE AND LUMBER STORAGE RACKS____ WORKBENCH WITH GAS OUTLETS____.

☐ MECHANICAL VENTILATION.

STAIRWAYS

Verify requirements and exit regulations with local code.

☐ TREADS AND RISERS, RISERS 7½" MAXIMUM, TREADS 10" MINIMUM____ NONSLIP TREADS____.

☐ STAIR WIDTH: OCCUPANT LOAD 50 OR FEWER, 36"; OCCUPANT LOAD GREATER THAN 50, 44". (Handrail, beam, or other projections may reduce required clearance width no more than 3½".)

☐ HANDRAILS AT EACH SIDE OF STAIR. (Maximum width of stair between intermediate handrails is 88".)

☐ HANDRAIL HEIGHT: 30" MINIMUM AND 34" MAXIMUM ABOVE TREAD NOSINGS.

☐ HANDRAIL ENDS: TERMINATE ON NEWELS OR RETURN TO WALL TO AVOID SLEEVE CATCHERS.

☐ POST POCKETS IN CONCRETE STAIRS LOCATED AND DESIGNED TO AVOID SPALLING OF CONCRETE.

☐ GUARDRAIL LOCATIONS: FLOOR OPENINGS____ OPEN LANDINGS____ GLAZED WALLS NEAR LANDINGS____.

☐ LANDINGS: MINIMUM ONE EACH 12' OF VERTICAL STAIR RUN____ MINIMUM LANDING WIDTH IN DIRECTION OF TRAVEL IS EQUAL TO STAIR WIDTH____.

☐ STAIRWELL LANDINGS: WALL CORNERS ROUNDED, ANGLED, OR WITH GUIDE RAILS TO AID FLOW OF EXIT TRAFFIC.

☐ BATTERY-POWERED EMERGENCY ILLUMINATION.

☐ INTERIOR STAIRWELLS WITH SKYLIGHTS: PROTECT SKYLIGHTS WITH WIRE MESH SCREENS BOTH ABOVE AND BELOW GLAZING.

☐ HEADROOM: 6'6" MINIMUM CLEARANCE BETWEEN NOSING AND SOFFIT.

☐ MAXIMUM DEVIATION IN RISER HEIGHTS AND TREAD WIDTH IN A SINGLE FLIGHT: ³⁄₁₆".

☐ UNDER-STAIR STORAGE AND ACCESS DOOR.

☐ UNDER-STAIR SOFFIT MATERIAL AND CONSTRUCTION.

☐ STAIR AND LANDING DIMENSIONS____ ELEVATION POINTS____ MATERIALS____ GUTTERS, DRAINS, AND GRATINGS WHERE STAIRS ARE EXPOSED TO WEATHER____.

REFLECTED CEILING PLANS

General reference data are the same as for Floor Plans.

☐ COLUMNS, POSTS, WALLS, AND CEILING-HIGH PARTITIONS.

☐ EXPOSED CEILING BEAMS, GIRDERS, OR JOISTS: LINE INDICATIONS____ MATERIALS AND FINISH____ SIZES____ SPACING____.

☐ SLOPE OF CEILINGS AND EXPOSED BEAMS: DIRECTIONS OF SLOPES____ LOW AND HIGH ELEVATION POINTS____.

☐ CONCEALED BEAMS, GIRDERS, AND HEADERS: BROKEN-LINE INDICATIONS____ MATERIALS____ SIZES____ SPACINGS____.

☐ HANGING, SURFACE-MOUNTED, AND RECESSED LIGHTING FIXTURES____ LIGHTING TRACKS____.

☐ SUSPENDED CEILING GRID, MAJOR AND SECONDARY CHANNELS.

☐ EGG CRATES____ LIGHT DIFFUSERS____.

☐ FIRE SPRINKLER SYSTEM (integrated with suspended ceiling, lighting, and partition systems).

☐ INTEGRATED CEILING RADIANT HEATING SYSTEM.

☐ FIRE-RESISTIVE CEILING CONSTRUCTION.

☐ DIVIDING LINES AND CEILING HEIGHTS AT CHANGES IN CEILING PLANE____ FURRED CEILINGS____.

☐ OUTLINE, NOTE, AND HEIGHTS: COVES____ SOFFITS____ LIGHTING VALANCES____ COFFERS____ TRACKS____ LIGHTING RECESSES____.

☐ CEILING TILE, PANELS, MIRRORS, BATTENS, AND TRIM: REFLECTED PLAN OR PARTIAL PLAN____ NOTES____ DIMENSIONS____.

☐ ACOUSTICAL TREATMENTS____ SOUND REFLECTORS____ REFLECTANCE BARRIERS____.

☐ HINGED OR REMOVABLE ACCESS PANELS____ FOLD-DOWN LADDERS____ SOFFIT SCUTTLES____.

☐ CATWALKS____ SERVICE LADDERS____.

☐ HATCHES____ FUSIBLE-LINK HATCHES____.

☐ THRU-CEILING SHAFTS____ CHASES____ CHUTES____ FUTURE SHAFTS____.

☐ SKYLIGHTS____ MONITORS____ LIGHT WELLS____ SHAFTS____ SKYLIGHT SUN SCREENS____.

☐ WALLS OR PARTITIONS EXTENDING THROUGH CEILING TO UNDERSIDE OF ROOF.

☐ ATTIC OR CEILING SPACE SEPARATORS____ DRAFT STOPS____ ATTIC SPRINKLERS____.

☐ NOISE BARRIERS AT SUSPENDED-CEILING SPACES AT LOW PARTITION LINES.

☐ LOUVERS____ GRILLES____ REGISTERS____.

☐ CEILING-MOUNTED DIFFUSERS____ RETURN AIR VENTS____ UNIT HEATERS____ UNIT VENTILATORS____ FANS____ EXHAUST VENTS AND FLUES____.

☐ SPEAKERS____ ALARMS____ CEILING-MOUNTED TV CAMERAS AND MONITORS____ MANUAL OR POWERED PROJECTION SCREENS____.

☐ NOISE- AND VIBRATION-DAMPENING CONNECTORS FOR FAN MOTORS AND OTHER NOISE-GENERATING, CEILING-MOUNTED EQUIPMENT.

☐ SMOKE AND HEAT DETECTORS____ ANEMOSTATS____.

☐ CEILING-MOUNTED TRACKS: CURTAINS AND DRAPES____ SLIDING DOORS____ FOLDING DOORS____ SCREENS____ MOVABLE PARTITIONS____.

☐ CEILING CRANES____ HOISTS____.

☐ HANGERS____ HOOKS____ RACKS____ CEILING-MOUNTED CASEWORK____.

☐ BEAM SOFFITS____ STAIR AND STAIR LANDING SOFFITS____.

☐ REFLECTED EXTERIOR ROOF OVERHANG SOFFIT PLAN: MESH OR LOUVER SOFFIT VENTS____ TRIM____ SCREEDS____ CONTROL JOINTS____.

ROOF PLAN

General reference data are the same as for Floor Plans.

☐ BUILDING OUTLINE AND ROOF OUTLINE. (Broken line building perimeter where shown below roof overhang.)

☐ OVERALL EXTERIOR WALL DIMENSIONS____ ROOF OVERHANG DIMENSIONS____.

☐ OUTLINE OF FUTURE ADDITIONS.

☐ CONNECTION WITH EXISTING STRUCTURES.

☐ BUILDING AND ROOF EAVE DIMENSIONS____ DIMENSIONS TO PROPERTY LINES____.

☐ BUILDING LOCATION REFERENCED TO BENCHMARK.

☐ REQUIRED PROPERTY SETBACK LINES.

☐ BAYS____ AREAWAYS____ BALCONIES____ LANDINGS____ DECKS____ STEPS____ SILLS____ OUTWARD-OPENING DOORS____ ALL PROJECTIONS LABELED AND DIMENSIONED SHOWING COMPLIANCE WITH PROPERTY LINE AND SETBACK REQUIREMENTS____.

☐ STAIR NUMBERS____ FIRE ESCAPES____ FIRE ESCAPE ROOF LADDERS____.

☐ ADJACENT AUXILIARY STRUCTURES.

☐ LOWER-STORY CANOPIES, MARQUEES: MATERIALS____ SLOPES____ SCUPPERS AND RAINWATER LEADERS____.

☐ ELECTRICAL-SERVICE ENTRY____ TELEPHONE ENTRY____ TV CABLE ENTRY____.

☐ OVERHEAD POWER, TV CABLE, AND PHONE LINES LOCATED TO CLEAR DRIVEWAYS OR WALKWAYS AT LEGALLY REQUIRED HEIGHTS.

☐ SUPPORTS AND GUYS FOR OVERHEAD CABLE____ PERISCOPE ENTRY HEADS____.

☐ ROOF FASCIA: MATERIALS____ CONSTRUCTION AND EXPANSION JOINTS____.

☐ CORNICE AND COPING BLOCK: MATERIALS____ EXPANSION JOINTS____.

☐ GRAVEL STOPS: INDICATION AND NOTES____ EXPANSION JOINTS____.

☐ ROOF GUTTERS: MATERIALS____ SIZES____ SLOPES AND DIRECTIONS OF SLOPES____ GUTTER SCREENS____ EXPANSION JOINTS____.

☐ DOWNSPOUTS AND LEADERS____ STRAINERS AT TOPS OF LEADERS____. (One sq. in. of downspout section per 150 sq. ft. of roof is typical.)

☐ RAIN DIVERTERS____ SCUPPERS____ SCUPPERS THROUGH PARAPETS____ OVERFLOWS____.

☐ CURBS AND PARAPETS: FLASHING INDICATION____ CANT STRIPS____ CRICKETS____ CAP FLASHING____.

☐ ROOF SHEATHING AND ROOF STRUCTURE SHOWN IN PARTIAL PLAN SECTION____ SIZES, MATERIALS, AND SPACING OF RAFTERS, ROOF JOISTS, OR TRUSSES____ PURLINS____ JOIST BRIDGING____.

☐ FINISH ROOFING MATERIALS: INDICATION AND NOTATION.

☐ ROOF EXPANSION AND CONSTRUCTION JOINTS WITH COVER PLATES.

☐ ROOF SLOPE: DIRECTION OF SLOPES AND DEGREE IN INCHES PER FOOT.

☐ VALLEYS____ HIPS____ RIDGES____.

☐ SKYLIGHTS, BUBBLES, OR MONITORS: LOCATIONS____ SIZES____ CURB AND CANT STRIP LINES____ MATERIALS____.

☐ WIRE MESH OVER SKYLIGHTS.

☐ LIGHT WELLS____ ATRIA____ COURTS____ SHAFTS____ SHAFT ROOFS____ SHAFT NUMBERS____.

☐ SOIL STACKS.

☐ AIR INTAKES____ EXHAUST VENTS____. (Air intakes located 10′ minimum distance from soil stack vents. Intakes and exhausts set at height to avoid clogging by snow.)

☐ SMOKE VENT HATCHES.

☐ ROOF-MOUNTED VENTILATORS____ EXPOSED DUCTWORK____.

☐ ATTIC RIDGE VENTS____ GABLE VENTS____ CUPOLAS____ DORMERS____ LOUVER AND/OR BIRD SCREEN SIZES AND MATERIALS____. (Typical vent area: 1 sq. in. for each 10 sq. ft. of attic area.)

☐ CHIMNEYS OR FLUES____ SADDLE AND FLASHING INDICATION AND NOTES____ CHIMNEY CAPS____ CEMENT WASH____.

☐ ELEVATOR PENTHOUSE____ ELEVATOR SHAFT VENT____ SMOKE VENT PENTHOUSES____ SMOKE VENT HATCHES____.

☐ STAIR BULKHEADS____ ROOF LADDERS____ ROOF SCUTTLES____.

☐ WINDOW-WASHING EQUIPMENT PENTHOUSE____ GATE AND PIPE RAIL FOR WINDOW-WASHING EQUIPMENT____.

☐ SUPPORTS____ ANCHORS____ MOUNTING POCKETS FOR ENTRY HEADS____ FLAGPOLES____ FLOODLIGHTING____ ANTENNA GUYS____ LIGHTNING RODS____ LOCATION DIMENSIONS____.

☐ COOLING TOWER PENTHOUSE____ COOLING TOWER EXHAUSTS LOCATED 100′ AWAY FROM ALL AIR INTAKES____.

☐ ROOF-MOUNTED SIGNS____ BROADCAST ANTENNA____ PARABOLIC REFLECTORS____ AIR-

CRAFT LIGHTS____ BUILDING IDENTIFICATION FOR POLICE HELICOPTERS____.

☐ ROOF-MOUNTED AIR-CONDITIONING EQUIPMENT____ REFRIGERATION EQUIPMENT____.

☐ FUTURE EQUIPMENT.

☐ HELIPORT: WINDSCREENS____ LIGHTS AND MARKERS____ LANDING PLATFORM____.

☐ HOSE BIBBS____ HYDRANTS____ STANDPIPES____ PIPING SUPPORTS____ STUB-UPS____.

☐ WATER TANKS.

☐ CONCRETE, TILE, OR WOOD DECK ON FLAT ROOFS: SIZE AND LOCATION DIMENSIONS____ MATERIAL INDICATIONS____.

☐ SOUND AND VIBRATION ISOLATION BETWEEN ROOFTOP EQUIPMENT ROOMS AND IMPORTANT ROOMS AT STORY BELOW ROOF.

☐ GUARDRAILS LOCATIONS: FLAT ROOF PERIMETERS____ LIGHT WELLS____ ATRIA____ COURTYARDS____ OPEN SHAFTS____.

☐ GUARDRAILS: SIZES____ MATERIALS____ HEIGHTS____.

☐ WEATHERPROOF ELECTRIC OUTLETS.

EXTERIOR ELEVATIONS

General reference data are the same as for Floor Plans.

SUBGRADE TO FLOOR LINE

☐ EXISTING AND NEW FINISH GRADE LINE AND ELEVATIONS (rough grade, finish grade, and topsoil depth sometimes shown)____ FILL AND ENGINEERED FILL____ CRUSHED-ROCK FILL____.

☐ EXISTING UNDERGROUND STRUCTURES: WHICH TO RETAIN____ WHICH TO REMOVE____.

☐ HOLES, TRENCHES, EXCAVATIONS TO BE FILLED.

☐ EXISTING ROCK OUTCROPS TO BE REMOVED____ ELEVATION POINTS____.

☐ ELEVATION POINTS AT BOTTOMS OF FOOTINGS.

☐ FOOTING, FOUNDATION WALL, AND BASEMENT WALL LINES BELOW GRADE—DASHED.

☐ BUILDING SLAB FOOTING LINES____ SLAB FLOOR LINE____ CHIMNEY FOOTINGS____ FOOTINGS AT OTHER CONCENTRATED LOADS____.

☐ DRAINPIPE AND GRAVEL BED AT FOOTINGS.

☐ BASEMENT WALL OPENINGS AND AREAWAYS____ AREAWAY GRILLE COVERS OR GUARDRAILS____.

☐ CRAWL SPACE VENTS AND ACCESS PANELS.

☐ SPLASH BLOCKS____ GRADE GUTTERS____ CURBS____ LEADER BOOTS____.

☐ EXTERIOR WALKS____ LANDINGS____ STEPS____ PAVING WITH ELEVATIONS AND SLOPES____.

☐ FOOTINGS AND ROCK SUB-BASE FOR EXTERIOR SLABS____ LANDINGS____ STEPS____.

☐ SOIL AND CRUSHED-ROCK INDICATIONS.

☐ DIMENSIONS: FOOTING THICKNESSES____ FOUNDATION WALL HEIGHTS____ DEPTHS OF AREAWAYS____ PAVING ELEVATIONS____ FLOOR ELEVATIONS____ BASEMENT FLOOR TO GROUND FLOOR DISTANCE____.

☐ CLEARANCES: CEMENT STUCCO OR MASONRY VENEER 8″ ABOVE FINISH GRADE____ STUCCO 12″ AWAY FROM SHRUBBERY____.

FLOOR TO CEILING OR ROOF LINE

☐ BUILDING WALL OUTLINE____ OFFSETS____ CANOPIES____ BALCONIES____.

☐ DIMENSIONS: FLOOR TO FLOOR HEIGHTS____ WINDOW, DOOR, OPENING HEAD HEIGHTS FROM FLOOR LINE____.

☐ NOTE WHETHER VERTICAL DIMENSIONS ARE TO FINISH SURFACES, SUBFLOORING, OR STRUCTURE____ WHETHER OPENINGS ARE ROUGH OR FINISH____.

☐ ADJACENT CURBS____ FENCING____ WALLS____ PLANTERS____ RETAINING WALLS____ PAVING____.

☐ WALL MATERIAL INDICATIONS, TEXTURES, PATTERNS, AND NOTES: BRICK OR BLOCK MASONRY COURSES AND HEADER COURSES____ STONE____ CONCRETE____ CEMENT PLASTER____ WOOD SIDING____ METAL SIDING____.

☐ WEEP HOLES: CAVITY WALLS____ MASONRY VENEER____ CURTAIN WALLS____.

☐ BATTENS: SIZES____ MATERIALS____ SPACING____.

☐ CONSTRUCTION JOINTS: EXPANSION AND CONTRACTION JOINTS____ CONTROL JOINTS IN MASONRY SPANDRELS OVER OPENINGS____ PLYWOOD SIDING MOVEMENT JOINTS (¼″ between panels)____.

☐ EXTERIOR STEPS____ STOOP, LANDING, AND HANDRAIL____ STOOP OR LANDING SLOPE AND ELEVATIONS (top of exterior landing 3″ below door threshold)____.

☐ DECK, PORCH, AND BALCONY FLOOR LINES____ SLOPE AND ELEVATION POINTS____ RAILINGS WITH NEWELS OR HORIZONTAL MEMBERS SPACED AT 9″ MAXIMUM____.

☐ EXTERIOR-MOUNTED AIR-CONDITIONING AND VENTILATING EQUIPMENT.

☐ POSTS, COLUMNS: MATERIALS____ SIZES____.

☐ BUMPERS, WHEEL GUARDS, CORNER GUARDS: MATERIALS____ SIZES____.

☐ EXTERIOR RAISED DECK FRAMING____ POSTS____ PIERS____ CROSS BRACING____ STRUCTURAL CONNECTIONS____.

☐ LOADING DOCK____ BUMPERS____ CANOPY HEIGHT CLEARANCE FOR TRUCKS____.

☐ WALL TRIM: MATERIALS____ SIZES____.

☐ LEDGERS: MATERIALS____ SIZES____.

☐ BARGEBOARDS: MATERIALS____ SIZES____.

☐ DOWNSPOUTS____ LEADERS____ LEADER CONNECTIONS TO WALL____.

☐ HOSE BIBBS____ HYDRANTS AT WALL____ STANDPIPES____ SIAMESE FIRE DEPARTMENT CONNECTIONS AND IDENTIFYING SIGN____.

☐ DOORS AND WINDOWS: SILLS____ TRIM____ MUNTINS____ MULLIONS____.

☐ OVERHEAD DOORS____ ROLLING DOORS____ ROLLING OR SLIDING GRILLES____.

☐ DOOR AND WINDOW SYMBOLS FOR SCHEDULE OF DOOR AND WINDOW TYPES AND SIZES.

☐ DOORS AND WINDOWS: DIRECTION OF SWING____ DIRECTION OF SLIDING UNITS____ SWING OF HOPPER UNITS____.

☐ CAULKING AT ALL SASH IN MASONRY WALLS____ DOOR AND WINDOW FLASHING____.

☐ SHUTTERS.

☐ FIXED GLASS AT NONOPERABLE WINDOWS____ JALOUSIES____.

☐ OBSCURE GLASS____ WIRE GLASS____ PLASTIC____.

☐ METAL OR MASONRY GRILLES AT WINDOWS (used mainly where windows are accessible from exterior walkways or stairs).

☐ THRU-WALL AIR INTAKES____ BIRD OR INSECT SCREENS____ LOUVERS____.

☐ EXTERIOR LIGHT FIXTURES____ WEATHERPROOF CONVENIENCE OUTLETS____.

☐ BUILDING NAME AND ADDRESS PLAQUES.

☐ BUILDING CORNERSTONE____ ARCHITECT'S PLAQUE____.

☐ NIGHT DEPOSITORY SLOT.

☐ EXTERIOR-MOUNTED ALARMS.

☐ WALL-MOUNTED FLAGPOLES.

☐ SIGNS.

☐ TRASH CAN ENCLOSURE.

☐ EXTERIOR GAS AND ELECTRIC METER ENCLOSURES____ EXTERIOR ELECTRIC SWITCH BOX ENCLOSURE____.

☐ FLASHING AT CHANGES IN BUILDING MATERIAL.

☐ DRIPS AT UNDERSIDE EDGE OF CANTILEVERED BALCONIES AND PROJECTING SILLS____ WINDOW SILL DRIPS____ SILL WASH-OFFS____.

☐ MARQUEES____ AWNINGS____ CANOPIES____ CANOPY SCUPPERS AND RAINWATER LEADERS____.

CEILING LINE TO ROOF

☐ FASCIAS: MATERIALS____ SIZES____.

☐ EXTERIOR SOFFIT LINES.

For other items, see the Floor Plans and Site Plans sections in this checklist.

- ☐ SOFFIT MATERIALS____ CONSTRUCTION AND CONTROL JOINTS____. (Cement plaster soffits and control joints at 12' spacing and at building and fascia intersections.)
- ☐ EAVE SOFFIT VENTS____ RIDGE VENTS____ ATTIC VENTS AND LOUVERS____ LOUVER SIZES AND MATERIALS____.
- ☐ ROOF GRAVEL STOP____ CAP FLASHING____.
- ☐ GUTTERS____ BOX GUTTERS____ RAIN DEFLECTORS____ DOWNSPOUTS____ LEADERS AND LEADER STRAPS____ SCUPPERS____.
- ☐ EAVE SNOW GUARDS.
- ☐ PARAPET CAP MATERIAL (parapet cap slope inward toward roof)____ EXPANSION JOINTS____.
- ☐ RAILINGS AT ROOF____ FIRE ESCAPE____ EXTERIOR ROOF ACCESS LADDER AND RAILINGS____.
- ☐ FINISH ROOFING MATERIALS.
- ☐ ROOF SLOPES.
- ☐ SKYLIGHTS, BUBBLES, AND MONITORS____ ROOF-MOUNTED MECHANICAL EQUIPMENT____ WATER TANK____ COOLING TOWER____ STAIR BULKHEADS____ ELEVATOR MACHINE ROOM____.
- ☐ CHIMNEY AND CHIMNEY FLASHING SADDLE____ CRICKET____ COPING____ CAP____ FLUE____ SPARK ARRESTOR (chimney coping block or cement wash slope toward flue)____.
- ☐ ANCHORS FOR TV, FM, OR SHORT-WAVE ANTENNAS____ LIGHTNING RODS____ WEATHER VANES____ OVERHEAD CABLE ENTRY HEADS____.

For other items, see Roof Plan.

BUILDING CROSS SECTIONS AND WALL SECTIONS

See Floor Plans, Reflected Ceiling Plan, Roof Plan, and Exterior Elevations in this checklist for items that may also appear in section drawings.

- ☐ EXISTING AND NEW FINISH GRADE LINES WITH ELEVATIONS.
- ☐ FOOTINGS AND FOUNDATION WALLS.
- ☐ DRAIN TILES____ ROOF DRAIN CONNECTIONS TO STORM SEWER____.
- ☐ SECTION AT BASEMENT AREAWAY: AREAWAY WALL AND FOOTING DIMENSIONS____ DRAIN____ WINDOW OR HATCH SECTION____.
- ☐ BASEMENT WALL MEMBRANE WATERPROOFING____ DETAIL FOR PIPE AND SLEEVES THAT PENETRATE WATERPROOFING____.
- ☐ BASEMENT OR FOUNDATION WALL GIRDER RECESS.
- ☐ ADJACENT STOOPS____ LANDINGS____ PAVING____.
- ☐ EXTERIOR SLAB FOOTINGS OR FROST CURBS____ TAMPED FILL____ SAND OR GRAVEL FILL____ VAPOR BARRIER____ REINFORCING____ SLAB THICKNESSES____.
- ☐ DOWELS CONNECTING SLABS TO FOOTINGS OR FOUNDATION WALLS.
- ☐ MASONRY WALL: TYPE____ THICKNESS____ MATERIAL INDICATIONS____ AIR SPACE AND WEEP HOLES____ COURSES AND HEADER COURSES____ SILL FLASHING____ HEAD FLASHING____ ANCHORS AND REINFORCING____ GROUTING____ PARGING____ CAULKING____.
- ☐ CURTAIN WALLS: SUPPORT ANGLES____ ANCHORS____ WEEP HOLES____ MOVEMENT JOINTS____ CAULKING____ SEALANTS____.
- ☐ GIRDERS, BEAMS, AND JOISTS____ HANGERS____ ANCHORS____ STIRRUPS____ POCKETS____ LEDGERS____ BEARING PLATES____ BRIDGING AND BLOCKING____.
- ☐ FLOOR-TO-FLOOR HEIGHTS____ SUBFLOORING____ FLOOR SLAB OR METAL DECK____ TOPPING____ FINISH FLOORING____.
- ☐ FLOOR PEDESTALS____ TRENCHES____ SINKAGES____ PITS AND GUARDRAILS____ CANTILEVERED SLABS____.
- ☐ FLOOR WATERPROOFING AT WET ROOMS.
- ☐ COLUMNS AND POSTS____ BASE PLATES____

☐ PIERS___ PEDESTALS___ FIREPROOFING___ COLUMN CAPS___ DROP PANELS___.

☐ COLUMN SPLICES___ STIFFENERS___ WIND BRACING___ CROSS TIES___.

☐ LINTELS___ HEADERS___ MASONRY LEDGES___ STEEL WALL ANGLES___ SPANDREL BEAMS___ SPANDREL ANGLES___.

☐ DOOR AND WINDOW: SILLS OR SADDLES___ HEAD FRAMING___ CAULKING___ FLASHING___ TRIM___ STOOLS AND APRONS___ ANCHORS TO WALL___.

☐ LIGHT COVES___ COFFERS___ VALANCES___.

☐ WALL AND CEILING VAPOR BARRIERS___ RIGID OR BATT THERMAL INSULATION___ FIREPROOF CONSTRUCTION___ SOUNDPROOFING___ ACOUSTICAL TREATMENTS___ WAINSCOTS___ FINISHES___.

☐ SUSPENDED CEILINGS___ CEILING-MOUNTED EQUIPMENT___ CEILING OR ATTIC FIRE BARRIERS___.

☐ MASONRY WALL TOP PLATE OR LEDGERS___ PLATE ANCHORS___ BEAM POCKETS___.

☐ PARAPET WATERPROOFING___ FLASHING REGLET___ FLASHING AND COUNTERFLASHING___ OVERFLOWS___.

☐ PARAPET CAP___ WASH TOWARD ROOF___ PARAPET EXPANSION AND CONSTRUCTION JOINTS___.

☐ EDGE FLASHING___ GUTTER AND GUTTER CONNECTORS___ GRAVEL STOP___ CAP FLASHING___ CANT STRIPS___ EAVE SNOW GUARDS___.

☐ TRUSSES___ TRUSS SPLICES___ SWAY BRACING___ SAG RODS___ PURLINS___.

☐ OVERHANG___ FASCIA___ SOFFIT___ SOFFIT VENTS___ TRIM___ EXPANSION JOINTS___.

☐ ROOF DECKING OR SHEATHING___ FINISH ROOFING MATERIALS___ ROOF EXPANSION JOINTS___ SLOPES___ THRU-ROOF VENT PIPE AND FLUE FLASHING___ CURBS___ ANCHORS___ PITCH POCKETS___ SNOW-MELTING EQUIPMENT___ RAILINGS___.

☐ THRU-BUILDING SHAFTS: ELEVATOR SHAFT, HOIST, AND PIT___ TRASH CHUTE, SPRINKLER, AND WASH-OUT SYSTEM___ CONVEYORS___ INCINERATOR___ PIPE AND ELECTRICAL CHASES___ INTERIOR RAINWATER LEADERS___.

☐ SKYLIGHTS___ LIGHT SHAFTS___ CONDENSATION COLLECTORS___ MESH GUARDS___.

SCHEDULES

DOOR SCHEDULE

☐ DOOR TYPES DRAWN IN ELEVATION (½" scale is typical).

☐ SYMBOL AND IDENTIFYING NUMBER AT EACH DOOR.

☐ DOOR WIDTHS___ HEIGHTS___ THICKNESSES___.

☐ DOOR TYPES___ MATERIALS___ FINISHES___. (Number of doors of each type and size is sometimes noted.)

☐ OPERATING TYPE: SLIDING___ SINGLE-ACTING___ DOUBLE-ACTING___ DUTCH___ PIVOT___ TWO-HINGE___ THREE-HINGE___.

☐ KICKPLATES.

☐ FIRE RATING IF REQUIRED.

☐ LOUVERS___ UNDERCUTS FOR VENTILATION___.

☐ SCREENS.

☐ MEETING STILES.

☐ DOOR GLAZING___ TRANSOMS___ BORROWED LIGHTS___.

☐ DETAIL KEY REFERENCE SYMBOLS: SILLS___ JAMBS___ HEADERS___.

☐ MANUFACTURERS___ CATALOG NUMBERS___. (If not covered in specifications.)

☐ METAL FRAMES: ELEVATIONS___ SCHEDULE___ DETAILS___.

WINDOW SCHEDULE

☐ WINDOW TYPES DRAWN IN ELEVATION (½" scale is typical).

☐ WINDOW SYMBOL AND IDENTIFYING NUMBER AT EACH DRAWING.

☐ WINDOW SIZE___ WIDTH___ HEIGHT___.

☐ WINDOW TYPE___ DIRECTION OF MOVEMENT OF OPERABLE SASH AS SEEN FROM EXTERIOR___. (Number of windows of each type and size is sometimes noted.)

☐ GLASS THICKNESS AND TYPE.

☐ SCREENS.

☐ NOTE: FIXED___ OBSCURE___ WIRE___ TEMPERED___ DOUBLE GLAZING___ TINTED___.

☐ DETAIL KEY REFERENCE SYMBOLS: SILLS___ JAMBS___ HEADERS___.

☐ MANUFACTURERS___ CATALOG NUMBERS___ (if not covered in specifications).

☐ METAL-FRAME STOREFRONT OR CURTAIN WALL SYSTEM: ELEVATIONS____ SCHEDULE____ DETAILS____.

FINISH SCHEDULE

☐ ROOM NAME AND/OR IDENTIFYING NUMBER.

☐ FLOOR: THICKNESS____ MATERIAL____ FINISH____.

☐ BASE: HEIGHT____ MATERIAL____ FINISH____.

☐ WALLS: MATERIALS____ FINISHES____. (Walls may be identified by compass direction if finishes vary wall by wall. Note waterproofing and waterproofing membrane wall construction.)

☐ WAINSCOT: HEIGHT____ MATERIAL____ FINISH____.

☐ CEILING: MATERIAL____ FINISH____.

☐ SOFFITS: MATERIAL____ FINISH____.

☐ CABINETS: MATERIAL SPECIES AND GRADE____ FINISH____.

☐ SHELVING: MATERIAL____ FINISH____.

☐ DOORS: MATERIAL____ FINISH____ (if not covered in Door Schedule).

☐ TRIM AND MILLWORK: MATERIAL SPECIES AND GRADE____ FINISH____.

☐ MISCELLANEOUS REMARKS OR NOTES.

☐ COLORS: STAIN AND PAINT____ MANUFACTURER AND TRADE NAMES OR NUMBERS (if not covered in specifications; may be left for later decision with provision for paint allowance by bidders)____.

☐ FINISHES: EXTERIOR WALLS____ SILLS____ TRIM____ POSTS____ GUTTERS AND LEADERS____ FLASHING AND VENTS____ FASCIAS____ RAILINGS____ DECKING____ SOFFITS____. (Included in finish schedule if not covered in specifications.)

ARCHITECTURAL WORKING-DRAWING CHECKLIST II: RESIDENTIAL AND OTHER LIGHT-FRAME CONSTUCTION

COVER SHEET AND TITLE BLOCKS

COVER SHEET

☐ PROJECT NAME____ PERSPECTIVE RENDERING OF PROJECT____. (Smaller projects usually have no title sheet; just start with site plan.)

☐ INDEX OF ARCHITECTURAL AND CONSULTANTS' DRAWINGS.

☐ GENERAL NOTES (or "General Conditions," if job doesn't include a separate set of specifications).

☐ LEXICON OF ABBREVIATIONS AND NOMENCLATURE.

☐ LEGEND OF MATERIALS INDICATIONS, SYMBOLS, AND DRAWING CONVENTIONS.

☐ LEGEND OF STRUCTURAL, ELECTRICAL, PLUMBING, AND HVAC SYMBOLS AND CONVENTIONS (if not shown on other drawings).

☐ LEGEND OF SITE WORK AND SURVEY DRAWING CONVENTIONS AND SYMBOLS (if not shown on site plan).

TITLE BLOCKS

☐ ARCHITECT'S NAME____ ADDRESS____ PHONE NUMBER____ REGISTRATION NUMBER OR OFFICIAL STAMP____.

☐ PROJECT NAME AND ADDRESS.

☐ OWNER'S NAME AND ADDRESS.

☐ CONSULTANTS' NAMES____ ADDRESSES____ PHONE NUMBERS____ REGISTRATION NUM-

BERS OR STAMPS____.

☐ DRAWING TITLE AND SCALE.

☐ MINI OUTLINE PLAN OF BUILDING (usually used for larger buildings).

☐ PROJECT ARCHITECT AND/OR JOB CAPTAIN.

☐ DESIGNER AND DRAFTING TECHNICIANS. (A growing preference is to provide full names of participants rather than initials.)

☐ INITIALS OR NAME OF DRAWING CHECKER.

☐ CHECKING AND PROGRESS PRINT DATES.

☐ SPACE FOR REVISION DATES AND REVISION REFERENCE SYMBOLS.

☐ PROJECT NUMBER____ FILE NUMBER____.

☐ SPACE FOR GENERAL NOTES____ PREPRINTED GENERAL NOTES____.

☐ COPYRIGHT NOTICE OR NOTE ON RIGHTS AND RESTRICTIONS OF OWNERSHIP AND USE OF DRAWINGS.

☐ SPACE FOR APPROVAL STAMPS OR INITIALS____ DATES OF APPROVALS____ BUILDING AND PLANNING AUTHORITY NAMES____.

☐ FINAL RELEASE DATE.

☐ DRAWING SHEET NUMBER AND TOTAL NUMBER OF DRAWINGS.

SITE PLAN

GENERAL REFERENCE DATA

☐ DRAWING TITLE AND SCALE (beneath drawing or on title block).

☐ ARROWS SHOWING COMPASS NORTH AND REFERENCE NORTH.

☐ PROPERTY ADDRESS: STREET, CITY, COUNTY, AND STATE.

☐ OWNER'S NAME____ ADDRESS____ (if not shown on title block).

☐ PROPERTY LEGAL DESCRIPTION: LOT____ BLOCK____ TRACT____ SUBDIVISION____ MAP BOOK____.

☐ LOT AND BLOCK NUMBERS OF ADJACENT PROPERTIES.

☐ SURVEYOR'S NAME____ ADDRESS____ PHONE NUMBER____ REGISTRATION NUMBER____.

☐ SURVEY DATE.

☐ LANDSCAPE CONSULTANT'S NAME____ ADDRESS____ PHONE NUMBER____.

☐ SOILS ENGINEER'S NAME____ ADDRESS____ PHONE NUMBER____ REGISTRATION NUMBER____.

☐ TEST-BORING CONTRACTOR'S NAME____ ADDRESS____ PHONE NUMBER____.

☐ SOIL TEST LAB'S NAME____ ADDRESS____ PHONE NUMBER____.

☐ SMALL-SCALE LOCATION OR VICINITY MAP SHOWING NEIGHBORING STREETS AND NEAREST MAJOR ARTERIES. (May be on cover sheet.)

☐ PROPERTY SIZE IN SQUARE FEET OR ACRES.

☐ NOTE REQUIRING BIDDERS TO VISIT SITE AND VERIFY CONDITIONS BEFORE SUBMITTING BIDS.

☐ LEGENDS OF SITE PLAN SYMBOLS AND MATERIALS INDICATIONS.

SITE DATA FROM PROPERTY SURVEY

The survey is verified by the architect's or designer's review of site conditions. The survey drawing is often copied and incorporated with the architectural site plan. If the survey drawing is complex, it may be included with the architectural drawings "as is" and followed by a simplified architectural site plan showing the building and new site work.

☐ PROPERTY BOUNDARY LINES____ DIMENSIONS____ DIRECTIONS OF PROPERTY LINES SHOWN IN DEGREES, MINUTES, AND SECONDS____.

☐ MONUMENT____ BENCHMARK____ REFERENCE POINTS____ STAKES____ REFERENCE DATUM____.

☐ CONTOUR LINES WITH GRADE ELEVATIONS. (Contour lines aren't shown on flat lots, but grade elevation points are noted at corners and at sloping portions of the property.)

☐ ± NOTES ON ELEVATION POINTS IF GRADES ARE BOTH ABOVE AND BELOW DATUM POINT.

☐ ON-SITE AND ADJACENT EASEMENTS____ RIGHTS-OF-WAY____. (Identities and dimensions.)

☐ REQUIRED FRONT, REAR, AND SIDE SETBACK LINES____ DIMENSIONS TO PROPERTY LINE____.

☐ ADJACENT STREETS OR ROADS: NAMES____ WIDTH DIMENSIONS____ ELEVATION POINTS TO STREET CENTERLINES____.

☐ ADJACENT STREETS OR ROADS TOTAL RIGHTS-OF-WAY____ EASEMENTS FOR ROAD WIDENING____.

☐ STREET STORM MAINS____ BASINS____ DIRECTION AND ELEVATION POINTS OF DRAIN SLOPES____.

☐ STREET SEWER HOOKUP____ SEWER MANHOLE____ ELEVATION POINT AT TOP OF MANHOLE____ ELEVATION POINT AT SEWER INVERT____ DIRECTION OF FLOW____.

☐ STREET GAS AND ELECTRIC MANHOLES____ ELECTRICAL VAULTS____ OTHER UTILITIES IN STREET OR ADJACENT TO STREET____.

☐ EXISTING CURBS AND WALKWAYS ADJACENT TO PROPERTY____ DIMENSIONS AND ELEVATION POINTS____.

☐ CURB ELEVATION POINTS SHOWN AT EXTENSIONS OF PROPERTY LINES.

☐ FIRE HYDRANTS____ TRAFFIC LIGHTS____ SIGNS____ PARKING METERS____.

☐ EXISTING SIDEWALK TREES OR SHRUBS.

☐ EXISTING POWER POLE(S).

☐ GAS MAIN AND HOOKUP LOCATION.

☐ STREET WATER AND HOOKUP LOCATION____ EXISTING WATER METER LOCATION____.

☐ OVERHEAD OR BURIED CABLE ON OR ADJACENT TO PROPERTY.

☐ EXISTING BUILDINGS ON ADJACENT PROPERTY WITHIN 5' OF PROPERTY LINE.

☐ ADJACENT BUILDING FOOTINGS AND FOUNDATION WALLS AT PROPERTY LINE.

☐ EXISTING ON-SITE PAVING, CURBS, AND WALKWAYS: DIMENSIONS____ MATERIALS____ ELEVATION POINTS____.

☐ EXISTING ON-SITE STRUCTURES: IDENTITY____ SIZE____ FLOOR ELEVATION POINTS____.

☐ EXISTING FENCES AND WALLS: MATERIALS____ HEIGHTS____.

☐ EXISTING UNDERGROUND STRUCTURES____ FOUNDATIONS____ HOLES AND TRENCHES____.

☐ EXISTING BURIED OR SURFACE TRASH.

☐ EXISTING ON-SITE UTILITIES____ SEWERS____ DRAINPIPE____.

☐ EXISTING ON-SITE OR ADJACENT DRAINAGE CULVERTS____ RIPRAP____.

☐ EXISTING SPRINGS OR WELLS.

☐ EXISTING PONDS____ CREEKS____ STREAMS____ OVERFLOW AREAS____.

☐ HIGH-WATER TABLE____ STANDING WATER____ MARSH____ QUICKSAND____.

☐ ROCK OUTCROPS: ELEVATION POINTS____ ROCK TO REMOVE____.

☐ EXISTING TREES: IDENTITIES____ TRUNK SIZES____ APPROXIMATE FOLIAGE AREA____. (Trees with trunks less than 6″ in diameter aren't usually noted. Architect may want smaller trees identified for removal and transplanting.)

☐ EXISTING MAJOR SHRUBS____ UNDERGROWTH____ GROUND COVER AREAS____.

SITE WORK AND APPURTENANCES

☐ NEW SITE CONTOURS____ NEW FINISH GRADES____ ELEVATION POINTS____.

☐ NEW BUILDING AND RELATED STRUCTURES____ OVERALL EXTERIOR WALL DIMENSIONS____ DIMENSIONS TO PROPERTY LINES____. (For other building components often included on site plan, see Roof Plan.)

☐ NEW BUILDING FINISH FLOOR ELEVATION AT GROUND FLOOR OR BASEMENT.

☐ NEW FINISH GRADE ELEVATIONS AT BUILDING CORNERS____ GRADE SLOPES AWAY FROM BUILDING AT ALL SIDES____.

☐ FOUNDATION OR BASEMENT EXCAVATION LIMIT LINES (usually 5' from building line).

☐ TEST-BORING LOCATIONS____ BORING TEST PROFILE FROM SOILS ENGINEER____. (May be separate drawing.)

☐ CUT AND FILL PROFILE. (May be separate drawing.)

☐ NEW BENCHMARK AND/OR BOUNDARY MARKERS.

☐ EXISTING HOLES AND TRENCHES: WHICH TO RETAIN____ WHICH TO FILL IN____.

☐ NEW FILL____ NOTE ON SOIL COMPACTION____.

☐ EXISTING FENCES AND WALLS: WHICH TO RETAIN____ WHICH TO REPAIR____ WHICH TO REMOVE____.

☐ EXISTING PAVING, WALKS, STEPS, AND CURBS: WHICH TO RETAIN____ WHICH TO REPAIR____ WHICH TO REMOVE____.

☐ EXISTING STRUCTURES: WHICH TO RETAIN____ WHICH TO REPAIR____ WHICH TO RELOCATE OR REMOVE____.

☐ OUTLINE OF FUTURE ADDITIONS.

☐ EXISTING ON-SITE UTILITIES, SEWERS, AND DRAINS: WHICH TO RETAIN____ WHICH TO USE____ WHICH TO REMOVE____.

☐ EXISTING TREES, SHRUBS, AND UNDERGROWTH: WHICH TO RETAIN____ WHICH TO REMOVE____ WHICH TO RELOCATE AND STORE FOR TRANSPLANTING____.

☐ TREES TO BE WRAPPED____ TREE GUARDS FOR PROTECTION DURING CONSTRUCTION____.

☐ AREAS FOR CLEARING AND GRUBBING____ EXISTING TRASH TO BE REMOVED____ STUMPS TO BE REMOVED____.

☐ TOPSOIL STORAGE AREA.

☐ TEMPORARY EROSION CONTROL.

☐ NEW RETAINING WALLS: MATERIAL____ HEIGHTS____ DIMENSIONS____ FOOTING LINES____ DRAINS____ DETAIL KEYS____.

☐ NEW SURFACE AND SUBSURFACE SOIL DRAINAGE____ DRY WELLS____.

☐ BUILDING PERIMETER FOUNDATION DRAIN TILE____ SLOPE AND DIRECTION OF DRAIN TO DRY WELL OR STORM SEWER____.

☐ BUILDING RAIN LEADER DRAINAGE____ SPLASH BLOCKS____.

☐ NEW FENCES, GATES, AND WALLS: MATERIALS____ HEIGHTS____ DIMENSIONS____ DETAIL KEYS____.

☐ NEW PUBLIC CURB AT STREET DRIVEWAY ENTRY.

☐ STREETSIDE MAILBOX____ ADDRESS PLAQUE____.

☐ DRIVEWAY GATE OR CHAIN____ HOUSE INTERCOM POST____ AUTOMATIC GATE____.

☐ NEW DRIVEWAY: MATERIAL____ DIMENSIONS____ CENTERLINE ELEVATION POINTS____ DRAINAGE SLOPES____ DRAINS____.

☐ CONSTRUCTION JOINTS IN CONCRETE PAVING. (Use of ½″ joints each 20′ to 30′ is typical, plus joints at connections with other construction.)

☐ DRIVEWAY AND PARKING AREA GRADES (Minimum 0.5%, maximum 5% at crown.)

☐ MANHOLES____ CLEANOUTS____ DRAINS____ GRATINGS IN PAVEMENT____.

☐ VEHICULAR WHEEL CURB, BUMPER, OR GUARD RAILING LOCATIONS: VULNERABLE WALLS____ LEDGES____ WALKWAYS____ TREES____ STANDARDS____ COLUMNS____.

☐ EXTERIOR PARKING LOCATED MINIMUM 5′ FROM WALLS AND OTHER STRUCTURES, AND LOCATED OUTSIDE TREE DRIP LINES AND ICICLE OR ROOF SNOW DROP LINES.

☐ PAVEMENT AND WALKWAY ICE-MELTING EQUIPMENT.

☐ NEW TERRACES____ PATIOS____ OFF-THE-GROUND DECKS____. (Terrace and patio with closed joint pavers or solid paving: minimum 1″ in 10′ slope for drainage.)

☐ TERRACE, PATIO, AND DECK: MATERIAL____ DIMENSIONS____ ELEVATION POINTS____ CONSTRUCTION NOTE____ DETAIL KEYS____.

☐ NEW WALKS AND STEPS: MATERIAL____ DIMENSIONS____ NOTES____ DETAIL KEYS____ ELEVATION POINTS____. (Typical small-building walks are 3′ to 5 ′ wide. No fewer than three risers at any point along walk or ramp. Handrail at steps over three risers. Rough, nonskid surfaces at exterior walks, steps, and landings. Maximum slope for walkway without handrails is 1 in 8.)

☐ WALKWAY LOW POINTS SLOPED TO DRAIN.

☐ EXTERIOR STAIR TREADS SLOPED TO DRAIN____ METAL TREAD NOSINGS AT CONCRETE STEPS____.

☐ CONCRETE WALK CONSTRUCTION JOINTS. (Use of ½″ joints each 30′, tooled joints at 5′ is typical. Construction joints at connections with other construction.)

☐ OPENINGS OR GRATINGS AT TREES IN PAVED AREAS FOR WATERING SPACE.

☐ CURBS OR BORDER BOARDS AT BARK, GRAVEL, AND PLANTING BEDS.

☐ NEW PLANTERS____ PLANT TUBS____ BENCHES____ VINE LATTICES____.

☐ NEW LANDSCAPE AREAS WITH LANDSCAPE LEGEND AND KEYS TO IDENTIFY TREE, PLANT, AND LAWN TYPES. (Extensive work may require separate landscaping plan by architect or consultant.)

☐ LANDSCAPE IRRIGATION AND/OR SPRINKLER SYSTEM: AUTOTIMER____ VALVES____ SPRINKLER HEAD LOCATIONS____. (May be shown on separate landscaping plan or plumbing plan.)

☐ DECORATIVE YARD LIGHTING AND TIMER SWITCH. (May be shown on electrical or landscaping plans.)

- ☐ RETAINING WALLS OR WELLS FOR EXISTING TREES AFFECTED BY CHANGES IN FINISH GRADE: MATERIAL____ DIMENSIONS____ NOTE____ DETAIL KEYS____.

- ☐ GAZEBO, GREENHOUSE, LATH HOUSE, STORAGE SHEDS: DIMENSIONS FOR LOCATION AND SIZE____ CONSTRUCTION NOTES____ DETAIL KEYS____.

- ☐ DECORATIVE POOL, FOUNTAIN, AND EQUIPMENT: MATERIALS____ UTILITIES____ DIMENSIONS____ NOTES____ DETAIL KEYS____.

- ☐ SWIMMING POOL, RELATED EQUIPMENT AND STRUCTURES: LOCATION AND SIZE DIMENSIONS____ MATERIALS____ UTILITIES____ CONSTRUCTION NOTES____ DETAIL KEYS____.

- ☐ HOSE BIBBS____ SILL COCKS____ HYDRANTS OR STUB-UPS____.

- ☐ WATER WELL AND PUMP HOUSING: DIMENSIONS FOR LOCATION AND SIZE____ CONSTRUCTION NOTES____ DETAIL KEY____.

- ☐ NEW WATER SUPPLY MAIN FOR HOUSE.

- ☐ NEW SEWER MAIN____ LOCATION OF VENT AND CLEANOUT____ DIRECTION OF SLOPE____ ELEVATION POINTS____.

- ☐ NEW GAS MAIN____ GAS METER____ SHUTOFF VALVE AT BUILDING____.

- ☐ NEW BURIED OR OVERHEAD ELECTRIC CABLE____ PHONE CABLE____ TV CABLE____.

- ☐ NEW TRENCHES: GAS____ WATER____ SEWER____ CABLE (minimum and maximum depth limits).

- ☐ TRENCH EXCAVATIONS LOCATED TO AVOID DAMAGE TO THREE ROOTS AND NEIGHBORING STRUCTURES.

- ☐ FUEL OIL OR LIQUIFIED GAS STORAGE TANK____ FUEL LINE TO BUILDING____ VENT____ GAUGE BOX____ FILL BOX____ MANHOLE____.

- ☐ PACKAGE SEWAGE TREATMENT TANK AND HOUSING: SIZE____ LOCATION____ CONSTRUCTION NOTE____ DETAIL KEYS____.

- ☐ SEPTIC TANK____ DISTRIBUTION BOX____ SEEPAGE PIT____ LEACHING FIELD____ SIZE AND LOCATION DIMENSIONS____ DETAIL KEYS____. (May be separate drawing.)

ROOF PLAN

The roof plan is commonly shown on the site plan of smaller buildings, but most of this checklist also applies if the roof plan is provided as a separate, non-site plan drawing.

- ☐ BUILDING OUTLINE AND ROOF OUTLINE. (Broken-line building perimeter where shown below roof overhang.)

- ☐ OVERALL EXTERIOR WALL DIMENSIONS____ ROOF OVERHANG DIMENSIONS____.

- ☐ OUTLINE OF FUTURE ADDITIONS.

- ☐ CONNECTION WITH EXISTING STRUCTURES: CONSTRUCTION NOTE____ DETAIL KEYS____.

- ☐ BUILDING AND ROOF EAVE DIMENSIONS____ DIMENSIONS TO PROPERTY LINES____.

- ☐ BUILDING LOCATION REFERENCED TO BENCHMARK.

- ☐ BAYS____ AREAWAYS____ BALCONIES____ LANDINGS____ DECKS____ STEPS____ SILLS____ OUTWARD-OPENING DOORS____ ALL PROJECTIONS LABELED AND DIMENSIONED SHOWING COMPLIANCE WITH PROPERTY LINE AND SETBACK REQUIREMENTS____.

- ☐ GROUND FLOOR ELEVATION.

- ☐ ADJACENT STRUCTURES, BREEZEWAYS, TRELLISES: DIMENSIONS____ NOTES____ DETAIL KEYS____.

- ☐ LANDINGS, STOOPS, TERRACES, WALKS: MATERIALS____ DIMENSIONS____ ELEVATIONS ____ NOTES____ DETAIL KEYS____.

- ☐ HOSE BIBBS AND HYDRANTS AT PERIMETER OF BUILDING. (May be shown on floor plan or separate plumbing plan.)

- ☐ ELECTRICAL-SERVICE ENTRY____ ELECTRIC METER____ SWITCH BOX____ TELEPHONE ENTRY____ TV CABLE ENTRY____.

- ☐ OVERHEAD POWER, TV CABLE, AND PHONE LINES LOCATED TO CLEAR DRIVEWAY OR WALKWAYS AT LEGALLY REQUIRED HEIGHTS.

- ☐ SUPPORTS AND GUYS FOR OVERHEAD CABLE____ PERISCOPE ENTRY HEADS____.

- ☐ EXTERIOR WALL CONVENIENCE OUTLETS AND LIGHTS. (Waterproof fixtures with ground fault detectors noted. May be shown separately on floor plans or electrical drawings.)

- ☐ ROOF FASCIA____ BARGEBOARDS AT GABLE____.

- ☐ CORNICE AND COPING BLOCK: MATERIALS____ JOINTS____ DIMENSIONS____ NOTES____ DETAIL KEYS____.

- ☐ GRAVEL STOP: INDICATION AND NOTE.

- ☐ ROOF GUTTER: MATERIAL____ SIZE____ DIRECTION OF SLOPE____ GUTTER SCREEN____.

- ☐ DOWNSPOUTS AND LEADERS____ STRAINERS AT TOPS OF LEADERS____. (one sq. in. of downspout section per 150 sq. ft. of roof is typical.)

- ☐ RAIN DIVERTERS____ SCUPPERS____ OVERFLOW SCUPPERS THROUGH PARAPETS____.

- ☐ CURBS AND PARAPETS: FLASHING INDICATION AND NOTES.

- ☐ ROOF SHEATHING AND ROOF STRUCTURE

SHOWN IN PARTIAL PLAN SECTION___ SIZES, MATERIALS, AND SPACING OF RAFTERS, ROOF JOISTS, OR TRUSSES___ PURLINS___ JOIST BRIDGING___.

☐ FINISH ROOFING MATERIAL(S): INDICATION AND NOTATION.

☐ ROOF SLOPE: DIRECTION OF SLOPES AND DEGREE IN INCHES PER FOOT.

☐ VALLEYS, HIPS, AND RIDGES. (Actual hip or gable framing length is sometimes shown in broken line and dimension is noted.)

☐ SKYLIGHTS, BUBBLES, OR MONITORS: LOCATIONS___ SIZES___ CURB LINES___ MATERIALS ___ NOTES___ DETAIL KEYS___.

☐ LIGHT WELLS___ ATRIA___ COURTS___.

☐ AIR INTAKES___ EXHAUST VENTS___. (Air intakes located 10' minimum distance from soil stack vents. Intakes and exhausts set at height to avoid clogging by snow.)

☐ ATTIC RIDGE VENT___ GABLE VENTS___ CUPOLAS___ DORMERS___ LOUVER AND/OR BIRD SCREEN SIZES AND MATERIALS___. (Typical vent area is 1 sq. in. for each 10 sq. ft. of attic floor area.)

☐ CHIMNEYS OR FLUES___ SADDLE AND FLASHING INDICATION AND NOTES___ CHIMNEY CAP___ CEMENT WASH___.

☐ SUPPORTS___ ANCHORS___ MOUNTING POCKETS FOR ENTRY HEADS___ FLOODLIGHTING___ ANTENNA GUYS___ LIGHTNING RODS___. LOCATION DIMENSIONS___ NOTES___ DETAIL KEYS___.

☐ ROOF-MOUNTED AIR-CONDITIONING EQUIPMENT.

☐ ROOF SCUTTLE FROM ATTIC.

☐ CONCRETE, TILE, OR WOOD DECK ON FLAT ROOFS: SIZE AND LOCATION DIMENSIONS___ MATERIAL INDICATIONS___ NOTES___ DETAIL KEYS___.

☐ GUARDRAILS AT ROOF DECKS: HEIGHT___ RAILING SIZES___ MATERIAL___ NOTES___ DETAIL KEYS___.

☐ SNOW-MELTING EQUIPMENT.

☐ EAVE SNOW GUARDS: CONSTRUCTION NOTES___ DETAIL KEYS___.

☐ OVERALL BUILDING CROSS-SECTION ARROWS___ DETAIL KEYS: WALL SECTION___ EAVE___ OVERHANG SOFFIT___.

☐ EXTERIOR ELEVATION VIEWS KEYED BY ARROWS AND IDENTIFICATION NUMBER. (Often just titled "front," "rear," etc., on exterior elevations of smaller buildings.)

FOUNDATION PLAN

CONCRETE SLAB ON GRADE

☐ SLAB PERIMETER OUTLINE AND DIMENSIONS.

☐ BROKEN LINES SHOWING PERIMETER FOOTINGS.

☐ INTERIOR FOOTINGS: PIERS___ COLUMNS___ BEARING WALLS___.

☐ FOOTING SIZE AND LOCATION DIMENSIONS.

☐ ELEVATION POINTS AT BOTTOMS OF FOOTINGS.

☐ EXISTING AND NEW FINISH GRADE LINES___ FINISH GRADE ELEVATION POINTS AT CORNERS OF BUILDING___.

☐ PERIMETER 4" DRAIN TILE WITH CRUSHED-ROCK BED.

☐ PERIMETER THERMAL INSULATION AT JUNCTURE OF SEPARATELY POURED SLAB AND FOOTING.

☐ PERIMETER MUD SILL___ SIZE, DEPTH, AND SPACING OF SILL ANCHOR BOLTS___. (Anchor bolts typically at 6'0" and 2'0" from corners. Bolts spaced so as not to be under studs.)

☐ GARAGE PERIMETER SLAB CURB: THICKNESS___ HEIGHT___.

☐ GARAGE SLAB SLOPED TO APRON___ CONSTRUCTION JOINT AT SLAB AND APRON___. (Typical slope is 2".)

☐ TERMITE SHIELDS.

☐ ADJACENT SLAB___ SIZE AND SPACING OF DOWEL CONNECTIONS BETWEEN SLABS___.

☐ FILLED, TAMPED, AND PUDDLED SOIL UNDER PORCH SLAB AND RAISED CONCRETE LANDINGS.

☐ PATIO OR TERRACE SLAB FROST CURB (6" edge of slab turned down 12" or more below finish grade).

☐ CONCRETE SLAB CONTROL JOINTS, CONSTRUCTION JOINTS. (Typical is ½" construction joints every 40' both ways. Tooled joints every 20' both ways and at piers and columns.)

☐ BROKEN-LINE OUTLINE OF FLOOR PLAN ON SLAB, INCLUDING PLUMBING FIXTURES

☐ CHANGES IN SLAB FLOOR ELEVATION.

☐ SLAB DEPRESSIONS FOR MASONRY, TILE, OR TERRAZZO FINISH FLOORING.

☐ SHOWER, SUNKEN TUB: SLAB DEPRESSIONS___ SIZES___ DEPTHS___ NOTES___ DETAIL KEYS___.

☐ FIREPLACE FOOTING: LOCATION AND SIZE DIMENSIONS____ REINFORCING NOTE____ BROKEN-LINE OUTLINE OF FIREPLACE ABOVE____.

☐ FIREPLACE ASHPIT AND CLEANOUT.

☐ UNDER-FLOOR OR PERIMETER DUCTWORK.

☐ RADIANT HEATING PIPE PLAN. (May be separate drawing.)

☐ CANTILEVERED SLAB SECTIONS____ CHANGE IN TENSILE REINFORCING FROM BOTTOM TO UPPER PORTION OF SLAB NOTED____.

☐ SLAB THICKNESS____ REINFORCING____ VAPOR BARRIER____ ROCK SUB-BASE____ SUBGRADE COMPACTION____.

☐ PARTIAL PLAN OF FINISH FLOOR: MATERIAL____ SLEEPERS____ PATTERN____ SLOPE____ THICKNESS OR HEIGHT OF FLOORING ABOVE SLAB____.

☐ DETAIL SECTION KEYS: FOOTINGS____ CURBS____ CONNECTIONS TO OTHER CONSTRUCTION____ POST AND COLUMN CONNECTIONS AT SLAB OR PIERS____.

☐ OVERALL BUILDING CROSS-SECTION ARROW AND REFERENCE SYMBOLS.

PERIMETER FOUNDATION, JOIST FRAMING

☐ PERIMETER FOUNDATION WALL AND FOOTING OUTLINE____ MATERIAL INDICATIONS____ WALL THICKNESSES____ FOOTING WIDTH____ OVERALL DIMENSIONS____ DETAIL KEYS____.

☐ FOUNDATION WALL LIP TO SUPPORT MASONRY VENEER____ NOTE AND DETAIL KEY____.

☐ STEPPED FOOTINGS AND FOUNDATION WALLS____ ELEVATION POINTS____ DETAIL KEYS____.

☐ ELEVATION POINTS AT BOTTOMS OF FOOTINGS. (Footing depths are typically below frost line and 12″ below undisturbed soil.)

☐ NOTE VERTICAL AND HORIZONTAL FOOTING AND FOUNDATION WALL REINFORCING. (May be shown separately in detail wall section.)

☐ EXISTING AND NEW FINISH GRADE LINES____ FINISH GRADE ELEVATION POINTS AT CORNERS OF BUILDING____.

☐ PERIMETER DRAIN TILE AND CRUSHED-ROCK BED.

☐ BROKEN-LINE OUTLINE OF FLOOR PLAN ABOVE, INCLUDING PLUMBING FIXTURES.

☐ GRADE BEAMS____ SIZE AND LOCATION DIMENSIONS____ DEPTH INTO SOIL____ REINFORCING____.

☐ INTERIOR FOUNDATION WALLS, DWARF WALLS, AND FOOTINGS AT BEARING WALLS WITHIN BUILDING____ SIZE AND LOCATION DIMENSIONS____.

☐ DECK SUPPORT PIERS AND FOOTINGS____ SIZE AND LOCATION DIMENSIONS____ MATERIALS____ DEPTH INTO SOIL____ DETAIL KEYS____ .

☐ GIRDER LINES____ GIRDER SIZE AND MATERIAL____ DETAIL KEYS____.

☐ GIRDER SUPPORT PIERS____ SIZE AND LOCATION DIMENSIONS____ WOOD CAPS OR POSTS AT CONCRETE PIERS____ DETAIL KEYS____.

☐ GIRDER SUPPORT RECESSES IN FOUNDATION WALLS (4″ bearing minimum, ½″ air space at ends and sides of wood girders)____ DETAIL KEYS____.

☐ FOUNDATION WALL MUD SILL: SIZE, DEPTH, AND SPACING OF ANCHOR BOLTS. (Anchor bolts typical at 6′0″, and 2′0″ from corners. Bolts spaced so as not to be under joists.)

☐ MUD SILL: SIZE____ MATERIAL____ GROUT LEVELING____.

☐ CRAWL SPACE ACCESS PANEL THROUGH FOUNDATION WALL. (A 24″ × 18″ opening is typical. Footing shown continuous below access panel.)

☐ CRAWL SPACE VENTILATION. (Use of 16″ × 8″ screened or louver vents between joists is typical. Usually 2 sq. ft. of vent is required for each 25 sq. ft. of crawl space area, with one vent within 3′ of each building corner.)

☐ CRAWL SPACE ACCESS THROUGH INTERIOR FOUNDATION WALLS.

☐ SCUTTLE TO CRAWL SPACE FROM FLOOR ABOVE.

☐ TERMITE SHIELDS.

☐ ADJACENT SLABS____ SIZE AND SPACING OF DOWEL CONNECTIONS BETWEEN SLABS, AND BETWEEN SLABS AND FOUNDATION WALL____.

☐ FIREPLACE SLAB____ FOOTING____ ASHPIT____ CLEANOUT____ SIZE AND LOCATION DIMENSIONS____ OUTLINE OF FIREPLACE ABOVE____ NOTE ON SLAB THICKNESS AND REINFORCING____ DETAIL KEY____.

☐ CONCRETE RODENT BARRIER OVER CRAWL SPACE SOIL.

☐ HERBICIDE AND/OR TERMITE SOIL TREATMENT. (May be covered in specifications).

☐ VAPOR BARRIER WITH SAND COVER OVER CRAWL SPACE SOIL (used mainly in wet-soil areas.)

☐ FLOOR JOISTS: SIZE____ TYPE____ MATERIAL____ SPACING____ ARROWS SHOWING DIRECTION OF SPANS____.

☐ JOIST BLOCKING OR BRIDGING: LOCATIONS____ MATERIAL____ SIZE____.

☐ DOUBLE JOISTS: UNDER PARALLEL PARTITIONS____ AT BATHTUBS____ AT OTHER CONCENTRATED LOADS____.

☐ DOUBLE HEADER JOISTS TO FRAME THRU-FLOOR OPENINGS.

☐ INDICATIONS, NOTES, AND DETAIL KEYS: FRAMING CLIPS____ ANCHORS____ BOLTS____ JOIST AND BEAM HANGERS____ POST AND GIRDER CONNECTIONS____ LEDGERS____.

☐ PARTIAL PLAN OF SUBFLOORING, SHEATHING, OR DECKING SHOWING MATERIAL AND PATTERN____ CONSTRUCTION NOTES____.

☐ FLOOR DIAPHRAGM SHEATHING WITH NAILING SCHEDULE AND CALCULATIONS.

☐ DETAIL REFERENCE KEYS: FOOTINGS____ FOUNDATION WALLS____ SLABS____ CURBS____ CONNECTIONS TO OTHER CONSTRUCTION____ FLASHING____.

☐ OVERALL BUILDING CROSS-SECTION ARROW AND REFERENCE SYMBOLS.

BASEMENT

The basement plan includes many elements also found on the previous foundation plan lists. Relevant elements are repeated here. All components normally shown in basement plans are included, although they do not pertain to foundation work or structure.

☐ BASEMENT WALL OUTLINE____ MATERIAL INDICATIONS____ WALL THICKNESS____ OVERALL DIMENSIONS____ DIMENSIONS OF OPENINGS____.

☐ BASEMENT WALL FOOTING OUTLINES IN BROKEN LINE____ WIDTH AND THICKNESS OF FOOTINGS____.

☐ ELEVATION POINTS AT BOTTOMS OF FOOTINGS____ CHANGES IN LEVEL OF FOOTINGS ____.

☐ NOTE VERTICAL AND HORIZONTAL FOOTING AND BASEMENT WALL REINFORCING (may be shown separately in detail wall section).

☐ BASEMENT WALL OPENINGS (DOORS, AREAWAYS, WINDOWS, VENTS): SILL HEIGHTS ABOVE FINISH FLOOR OR ELEVATION POINTS NOTED____ SCHEDULE AND DETAIL REFERENCE KEYS____.

☐ AREAWAYS____ SIZE AND LOCATION DIMENSIONS____ AREAWAY FLOOR ELEVATION POINTS____ SLOPE____ DRAINAGE____.

☐ SLEEVES FOR HOUSE SEWER MAIN, GAS, WATER, AND ELECTRIC SUPPLY: NOTED____

SIZED____ LOCATION AND HEIGHTS DIMENSIONED____ DETAIL KEYS AT SLEEVES AND PIPES THAT PASS THROUGH WATERPROOFING MEMBRANE____.

☐ PERIMETER BASEMENT WALLS: WATERPROOFING____ CAULKING____ INSULATION____ PARGING____ FURRING____ INTERIOR FINISH____ DETAIL KEYS____.

☐ EXISTING AND NEW FINISH GRADE LINES____ FINISH GRADE ELEVATION POINTS AT CORNERS OF BUILDING____.

☐ FINISH FLOOR ELEVATION POINTS____ FLOOR SLOPES____.

☐ PERIMETER DRAIN TILE AND CRUSHED-ROCK BED____ SLOPE____.

☐ PERIMETER MUD SILL ANCHOR BOLTS ATOP BASEMENT WALL: SIZE____ DEPTH____ SPACING____.

☐ MUD SILL SIZE AND MATERIAL____ NOTE GROUT LEVELING____.

☐ TERMITE SHIELDS.

☐ BROKEN-LINE OUTLINE OF FLOOR PLAN ABOVE, INCLUDING PLUMBING FIXTURES.

☐ UNEXCAVATED FLOOR AREA WITH ACCESS SCUTTLE AND FOUNDATION VENTS____ SIZES AND LOCATIONS OF SCUTTLE AND VENTS____.

☐ CONCRETE RATPROOFING OVER SOIL____ VAPOR BARRIER AT UNEXCAVATED FLOOR AREAS____.

☐ ADJACENT SLABS: FOOTINGS____ DIMENSIONS____ MATERIAL INDICATIONS____ SIZE AND SPACING OF DOWEL CONNECTIONS____ REINFORCING____ SOIL PREPARATION____.

☐ BASEMENT FLOOR SLAB: THICKNESS____ REINFORCING____ SUB-BASE____ WATERPROOFING____.

☐ BASEMENT FINISH FLOOR MATERIAL.

☐ FLOOR DRAINS____ SUMP PIT AND PUMP____. (Water supply to keep drain traps full during extended periods of nonuse.)

☐ FLOOR SLAB CONSTRUCTION JOINTS____ TOOLED JOINTS____ CAULKING AT JUNCTURE OF SLAB AND WALL____ CONSTRUCTION JOINT DETAIL KEYS____.

☐ CONTROL JOINTS IN SLAB AT PIERS AND COLUMNS.

☐ FLOOR ANCHORS____ PEDESTALS____ CURBS____ SLAB DEPRESSIONS____ CHANGES IN FLOOR LEVEL____.

☐ STEPS AND STAIRS: NUMBER OF RISERS____ STAIR WIDTH____ HANDRAIL____ MATERIALS ____ DETAIL OR STAIR SECTION KEYS____.

☐ COLUMNS AND POSTS____ SIZE____ MATERI-

□ AL___ FOOTING SIZES AND DEPTHS___ FOOTING CAPS OR BASE PLATES___ CONNECTION DETAIL KEYS___ CENTERLINE DIMENSIONS___.

□ STEEL PIPE COLUMNS___ SIZE___ BASE PLATE___ NONSHRINK CEMENT GROUT BASE___.

□ WOOD COLUMN AND POST SUPPORT PIERS. (Minimum of 2″ above floor. Noncorrosive metal barrier between pier and bottom of untreated wood.)

□ CONCRETE OR MASONRY PILASTERS___ SIZES AND CENTERLINE DIMENSIONS___. (Minimum 4″ × 12″ for girder bearing.)

□ GIRDER AND BEAM LINES___ TYPE___ MATERIAL___ SIZES___.

□ GIRDER SUPPORT RECESSES IN WALLS (Minimum of 4″ bearing, ½″ air space at ends and sides of wood girders)___ DETAIL KEYS___.

□ WOOD BEAMS, GIRDERS, AND JOISTS EDGE-BEVELED WHERE THEY ENTER MASONRY OR CONCRETE, SO TOP EDGE OF MEMBER IS 1″ MAXIMUM INTO WALL RECESS.

□ METAL WALL BOXES FOR WOOD BEAMS, GIRDERS, AND JOISTS ENTERING MASONRY OR CONCRETE WALL BELOW GROUND LEVEL___ WOOD PRESERVATIVE TREATMENT___.

□ FIREPLACE FOUNDATION AND FOOTING___ ASH PIT___ CLEANOUT___ SIZE AND LOCATION DIMENSIONS___ OUTLINE OF FIREPLACE ABOVE___ NOTE ON SLAB THICKNESS AND REINFORCING___ DETAIL OR FIREPLACE SECTION KEYS___.

□ OPENINGS IN BASEMENT CEILING FRAMING___ DUMBWAITER___ LAUNDRY CHUTE___ CHASES___ DUCTWORK___.

□ FLOOR JOISTS: SIZE___ TYPE___ MATERIAL___ SPACING___ ARROWS SHOWING DIRECTIONS OF SPANS___.

□ JOIST BLOCKING OR BRIDGING: SIZES___ LOCATIONS___ MATERIAL___.

□ JOIST LEDGERS: SIZE___ MATERIAL___ SIZE AND SPACING OF BOLTS TO WALL___.

□ DOUBLE JOISTS LOCATED UNDER PARALLEL PARTITIONS, BATHTUBS, AND OTHER CONCENTRATED LOADS.

□ DOUBLE HEADER JOISTS TO FRAME THRU-FLOOR OPENINGS.

□ INDICATIONS, NOTES, AND DETAIL KEYS: FRAMING CLIPS___ ANCHORS___ BOLTS___ JOISTS AND BEAM HANGERS___ POST AND GIRDER CONNECTIONS___.

□ PARTIAL PLAN OF UPPER FLOOR SUBFLOORING, SHEATHING, OR DECKING, SHOWING PATTERN AND MATERIAL___ CONSTRUCTION NOTES___.

□ COAL BIN AND CHUTE.

□ SPACE FOR FIXTURES OR EQUIPMENT PROVIDED BY OWNER___ N.I.C. NOTES___.

□ ELECTRICAL: SEE ELECTRICAL PLAN.

□ HOSE BIBBS, SINKS, OTHER PLUMBING: SEE PLUMBING PLAN.

□ FURNACE AND AIR-CONDITIONING EQUIPMENT: SEE FURNACE ROOM.

□ DOOR AND WINDOW SCHEDULE SYMBOLS.

□ INTERIOR ELEVATION ARROW SYMBOLS AND REFERENCE NUMBERS.

□ DETAIL REFERENCE KEYS: FOOTINGS___ FOUNDATION WALLS___ WATERPROOFING___ SLABS___ CURBS___ CONNECTIONS TO OTHER CONSTRUCTION___ FLASHING___.

□ OVERALL BUILDING CROSS-SECTION ARROW AND REFERENCE SYMBOLS.

FLOOR PLAN

WALLS AND PARTITIONS

☐ EXTERIOR WALLS____ COLUMNS AND POSTS____ WALL OPENINGS____. (Air space of 1″ between masonry veneer and wood sheathing.)

☐ OUTLINE OF FUTURE BUILDING ADDITIONS.

☐ DIMENSIONING (see next section on dimensioning).

☐ INTERIOR WALLS____ PARTITIONS____ COLUMNS AND POSTS____.

☐ OUTLINE AND NOTED HEIGHTS OF LOW PARTITIONS____ SCREENS____ PREFAB STORAGE WALLS____ DIVIDERS____ PLANTER WALLS____.

☐ OUTLINE OF CONSTRUCTED VENEERS____ MATERIALS____ HEIGHT OF VENEERS____ PANELING____ DADOS____ WAINSCOTS____ TILE WORK____.

☐ BROKEN-LINE OUTLINES: OVERHANGS____ CANOPIES____ OVERHEAD BALCONIES____ COVES____ VALANCES____ OTHER PROJECTIONS____. (May be on reflected ceiling plan.)

☐ DASHED WALL LINES ACROSS FRAMED OPENINGS.

☐ WALL OUTLINES: NOTE WHETHER FINISH SURFACES, ROUGH SURFACES, OR LINES OF FRAMING ARE REPRESENTED.

☐ NOTES AND DIMENSIONS OF FURRED WALLS____ PIPE AND VENT CHASES____. (Heights of chases noted if not shown in interior elevation drawings.)

☐ HEIGHTS AND SIZES OF SLEEVES AND OTHER THRU-WALL OPENINGS.

☐ SOUND-ISOLATION WALLS____ PARTY WALLS____.

☐ NOTES AND INDICATION OF WALLS WITH MEMBRANE WATERPROOFING.

☐ NOTES AND INDICATION OF FIRE-RESISTIVE CONSTRUCTION.

☐ MATERIALS HATCHING OR POCHÉ.

☐ PROTECTIVE RAILS, BUMPERS, OR CORNER GUARDS AT WALLS, GLAZING, COLUMNS, OR DOOR JAMBS SUBJECT TO DAMAGE.

☐ WALL-MOUNTED SHELVES____ CASEWORK____ MIRRORS____ PEGBOARDS____ TACKBOARDS____ HOOKS____ HANGERS____ RACKS____.

DIMENSIONING

☐ NOTES ON DIMENSIONING SYSTEM: MODULAR DIMENSIONING PRACTICE AND CONVENTIONS____ WHETHER DIMENSIONS ARE TO FINISH SURFACES, TO ROUGH SURFACES, OR TO ROUGH FRAMING____ NOTES OF SPECIAL ARROWHEAD INDICATIONS IF DIMENSIONING SYSTEM IS MIXED____.

☐ FINISH GRADE ELEVATION POINTS AT CORNERS OF BUILDING.

☐ OVERALL EXTERIOR WALL LENGTHS, OFFSETS, PROJECTIONS, AND RECESSES.

☐ EXTERIOR OPENINGS: DIMENSIONED TO FRAMES IN WOOD CONSTRUCTION____ DIMENSIONED TO CENTERLINES IN MASONRY CONSTRUCTION____. (Openings that clearly center on interior spaces or are adjacent to wall intersections often don't require dimensioning; door and window sizes may be covered in schedules.)

☐ INTERIOR WALL AND PARTITION LENGTHS____ DISTANCES BETWEEN INTERIOR WALLS OR PARTITIONS IN DIMENSION STRINGS ACROSS LENGTH AND WIDTH OF BUILDING____. (Dimension strings are located so as not to conflict with other essential notes, such as room names or numbers. Parallel strings of dimensions on interiors aren't considered desirable.)

☐ COLUMN, POST, AND MULLION CENTERLINE DIMENSIONS.

☐ DIMENSIONS OF OVERHANGS AND OVERHEAD PROJECTIONS.

☐ THICKNESSES OF EXTERIOR WALLS AND INTERIOR PARTITIONS. (Interior partitions are often dimensioned to centerlines, and wall thicknesses are usually not dimensioned in small-building construction.)

☐ DIMENSIONS OR DIMENSION NOTES OF CHASES AND FURRED WALLS.

☐ SMALL THRU-WALL OPENINGS____ NOTE LENGTH AND HEIGHT DIMENSIONS____ HEIGHT OF SILLS OF OPENINGS ABOVE FLOOR____.

☐ BUILT-INS____ SIZES____ DIMENSIONS TO FINISHED SURFACES____.

☐ RADII OF ARCS WITH REFERENCE DIMENSIONS TO LOCATE CENTERS OF ARCS OR CIRCLES.

☐ ANGLES IN DEGREES OF NON-RIGHT-ANGLE WALLS AND BUILT-INS. (In lieu of degree measurements, ends of angled unit or wall may be dimensioned from reference line or wall.)

WINDOWS

- ☐ GLASS LINES____ SILL AND STOOL LINES____ MULLIONS____.

- ☐ OPERABLE DIRECTION OF HORIZONTAL SLIDING WINDOWS____ SWINGS FOR VERTICAL CASEMENTS____ PIVOT WINDOWS____ AWNING WINDOWS____.

- ☐ HEADERS OR LINTELS: MATERIALS____ SIZES____. (Unless noted in separate structural drawings.)

- ☐ WINDOW SYMBOLS KEYED TO WINDOW SCHEDULE. (Window type, frame material, and size sometimes noted on plans of smaller buildings where separate window schedule is not included.)

- ☐ FIXED GLASS NOTED. (Interior and exterior fixed glass provided with removable stops on one side.)

- ☐ OBSCURE GLASS____ WIRE GLASS____ PLASTIC____ SAFETY GLASS____ TEMPERED GLASS____ DOUBLE GLAZING____. (May be noted only in window schedule.)

- ☐ EXTERIOR SHUTTERS____ GRILLE WORK____ SUN-CONTROL DEVICES____.

- ☐ WINDOW DETAIL REFERENCE KEYS (if not covered at window elevation drawings with schedule).

- ☐ INTERIOR DRAPE OR WINDOW SHADE TRACKS____ FOLDING SHUTTERS OR SCREENS____ VENETIAN BLIND POCKET____.

- ☐ WINDOW JAMBS: DESIGNED AND LOCATED TO AVOID CONFLICT OF MOLDING, TRIM, SILLS, OR STOOLS WITH ADJACENT WALLS OR DOORS.

- ☐ RESIDENTIAL FENESTRATION RULE: NATURAL LIGHT AREA EQUAL TO ONE-EIGHTH OF FLOOR AREA, MINIMUM 12 SQ. FT.____ NATURAL VENTILATION EQUAL TO ONE-SIXTEENTH OF FLOOR AREA IF MECHANICAL VENTILATION IS NOT PROVIDED____.

DOORS

- ☐ DOOR OPENINGS____ DOOR SWING DIRECTIONS____ SYMBOL KEY REFERENCES TO DOOR SCHEDULE____ NOTED SOLID-CORE EXTERIOR DOORS____ HOLLOW-CORE INTERIOR DOORS____ DOOR THICKNESSES ____ VENEER NOTES____. (In smaller buildings, door types and sizes are noted at each door without reference to schedule. Generous door sizes are recommended.)

- ☐ SLIDING AND HINGED SCREEN DOORS (shown in broken lines).

- ☐ LARGE DOORS: HEADER OR LINTEL MATERIALS AND SIZES.

- ☐ VERTICAL LATCH FOR FRENCH DOORS.

- ☐ REMOVABLE STILE FOR EXTERIOR DOUBLE DOORS.

- ☐ DOOR JAMBS: DESIGNED AND LOCATED TO AVOID CONFLICT OF MOLDINGS OR TRIM THAT INTERSECTS ADJACENT WALLS, DOORS, OR WINDOWS.

- ☐ DOOR OPENINGS: LOCATED TO AVOID TANGLING WITH NEIGHBORING DOORS IN ADJACENT WALLS.

- ☐ DOORS OPENING ON OPPOSITE SIDE OF COMMON CORRIDOR STAGGERED IN LOCATION FOR VISUAL AND SOUND PRIVACY.

- ☐ DOORS OPENING TO BATHROOMS, DRESSING ROOMS, OTHER PRIVATE AREAS LOCATED AND HINGED TO BLOCK DIRECT VIEW INTO ROOMS.

- ☐ SADDLE, JAMB, OR BUCK DETAIL KEYS OR NOTES IF NOT SHOWN ON SCHEDULE. (Metal bucks are sometimes indicated at door jambs on plan.)

- ☐ SADDLES OR SILLS____ WEATHERSTRIPPING AT EXTERIOR DOORS____ SADDLE MATERIAL____.

- ☐ DOOR UNDERCUTS____ LOUVERS____ KICKPLATES____ VIEW PANELS____ CLOSERS____ MIRRORS____. (Some of these may be covered in specifications or schedules.)

- ☐ FLOOR CUTS FOR PIVOT-HINGE DOORS____ FLOOR-MOUNTED DOORSTOPS____.

- ☐ DOOR ROLLER BUMPERS WHERE ADJACENT DOORS MAY HIT EACH OTHER.

- ☐ FLOOR-MOUNTED TRACKS FOR SLIDING DOORS____ SHOJI SCREENS____ SLIDING PANELS____ BI-FOLD PANELS____.

- ☐ OVERHEAD TRACKS FOR FOLDING AND SLIDING DOORS. (Sometimes specified on plans of smaller buildings.)

- ☐ DIRECTION OF MOVEMENT: SLIDING____ FOLDING____ BI-FOLD____ DOUBLE-ACTION____.

- ☐ SLIDING DOOR POCKETS. (Avoid conflict with wall switches and convenience outlets.)

- ☐ OVERHEAD FOLD-DOWN, ROLLING, OR SECTIONAL DOOR AT GARAGE____ AUTOMATIC DOOR OPENER____.

- ☐ DOUBLE-ACTING GATE AT KITCHEN/DINING ROOMS____ DUTCH DOOR____.

FLOORS

- ☐ FINISH FLOOR MATERIAL INDICATION AND/OR NAME. (Usually noted by type under each room name and described in detail in finish schedule and/or specifications.)

- ☐ DIVIDING LINES AND FLOOR HEIGHT NOTES AT CHANGES IN FLOOR LEVEL. (Note whether floor elevations apply to subfloor or finish floor.)

- [ ] STEPS: DIRECTION ARROW___ NUMBER OF RISERS___ HANDRAIL___.
- [ ] RECESSED SUBFLOORING FOR TILE, MASONRY, OR TERRAZZO FOR ALIGNMENT OF DIFFERENT ADJACENT FINISH FLOOR SURFACES.
- [ ] PATTERN OF MASONRY, TILE, MARBLE, OR TERRAZZO FLOORS. (Smooth surfaces should be avoided at landings, entries, and vestibules.)
- [ ] ROUGH SURFACES NOTED AT ENTRY LANDINGS, EXTERIOR STEPS.
- [ ] EXPANSION JOINTS, TOOLED JOINTS, CONSTRUCTION JOINTS, AND PERIMETER JOINTS IN SLABS___ JOINT DETAIL KEYS___.
- [ ] EXPANSION SPACE AT PERIMETER OF WOOD STRIP OR WOOD BLOCK FINISH FLOORING.
- [ ] SADDLES, THRESHOLDS, METAL STRIPS SEPARATING ONE FLOOR MATERIAL FROM ANOTHER.
- [ ] UNDERLAYMENT FOR RESILIENT FLOORS___ WATER- AND MILDEW-RESISTIVE UNDERLAYMENT BENEATH CARPET EXPOSED TO WATER, AS IN KITCHENS AND BATHS___.
- [ ] BUILT-IN EQUIPMENT AND BUILT-IN FLOOR-MOUNTED FURNISHINGS: FLOOR ANCHORS___ VIBRATION PADS___ CURBS___ PEDESTALS___.
- [ ] BALCONY, EXTERIOR LANDING, DECK: FLOOR MATERIAL___ FLOOR ELEVATION POINTS___ DIRECTION OF SLOPE___ LANDING AND BALCONY FLOOR DRAINS___ EXTERIOR SURFACE ELEVATION 3″ BELOW DOOR THRESHOLDS___.
- [ ] FLOOR OPENINGS FOR UTILITIES ACCESS PANELS.
- [ ] WATERPROOF-MEMBRANE CONSTRUCTION AT BUILT-IN SHOWERS AND TUBS.

REFLECTED CEILING PLAN

These data are usually provided on floor plans of smaller buildings without a separate reflected ceiling plan.

- [ ] CEILING JOISTS OR COMBINED CEILING AND ROOF JOISTS: DIRECTION OF SPAN___ JOIST MATERIAL___ SIZE___ SPACING___ JOIST BLOCKING/BRIDGING NOTED___.
- [ ] HEADERS AND LINTELS OVER LARGE OPENINGS___ MATERIALS AND SIZES___ DETAIL KEYS___.
- [ ] EXPOSED CEILING BEAMS, GIRDERS, OR JOISTS: LINE INDICATIONS___ MATERIALS AND FINISH___ SIZES___ SPACING___ DETAIL KEYS___.
- [ ] SLOPE OF CEILINGS AND EXPOSED BEAMS:

DIRECTIONS OF SLOPES___ LOW AND HIGH ELEVATION POINTS___.
- [ ] CONCEALED BEAMS AND GIRDERS: BROKEN-LINE INDICATIONS___ MATERIALS___ SIZES___ SPACING___ CONNECTION DETAIL REFERENCE KEYS___.
- [ ] POST AND BEAM CONNECTION DETAIL KEYS.
- [ ] DIVIDING LINES AND CEILING HEIGHTS AT CHANGES IN CEILING PLANE___ FURRED CEILINGS___ DETAIL KEYS___.
- [ ] OUTLINE, NOTE, AND HEIGHTS: COVES___ SOFFITS___ LIGHTING VALANCES___ COFFERS___ TRACKS___ LIGHTING RECESSES___.
- [ ] CEILING TILE, PANELS, MIRRORS, BATTENS, AND TRIM: REFLECTED PLAN OR PARTIAL PLAN___ NOTES___ DIMENSIONS___ DETAIL KEYS___.
- [ ] ATTIC ACCESS PANELS___ CONCEALED FOLD-DOWN LADDERS___ SOFFIT SCUTTLES___. (Attic access panels usually minimum 22″ × 30″.)
- [ ] SKYLIGHTS___ MONITORS___ LIGHT WELLS___ SHAFTS___.
- [ ] WALLS OR PARTITIONS EXTENDING THROUGH CEILING TO UNDERSIDE OF ROOF.
- [ ] CEILING-MOUNTED DIFFUSERS___ RETURN AIR VENTS___ UNIT HEATERS___ UNIT VENTILATORS___ FANS___ EXHAUST VENTS AND FLUES___.
- [ ] HANGING, SURFACE-MOUNTED, AND RECESSED LIGHTING FIXTURES___ SMOKE AND HEAT DETECTORS___. (For other items, see electrical plan.)
- [ ] CEILING-MOUNTED TRACKS: CURTAINS AND DRAPES___ SLIDING DOORS___ FOLDING DOORS___ SCREENS___ MOVABLE PARTITIONS___.
- [ ] HANGERS___ HOOKS___ RACKS___ CEILING-MOUNTED CASEWORK___.
- [ ] REFLECTED EXTERIOR ROOF OVERHANG SOFFIT PLAN: SCREEN OR LOUVER SOFFIT VENTS___ TRIM___ SCREEDS___ CONTROL JOINTS___ DETAIL KEYS___.

GENERAL REFERENCE DATA

- [ ] DRAWING TITLE AND SCALE.
- [ ] ARROWS SHOWING COMPASS NORTH AND REFERENCE NORTH.
- [ ] SQUARE FOOTAGE TOTALS: BUILDING___ AUXILIARY STRUCTURES___ DECKS AND BALCONIES___.
- [ ] ROOM NAMES OR NUMBERS___ EXTERIOR AREAS___.

☐ NOTE OF OVERALL ROOM SIZES (usually shown under each room name).

☐ INTERIOR ELEVATION ARROW SYMBOLS AND REFERENCE NUMBERS.

☐ EXTERIOR ELEVATION ARROW SYMBOLS AND REFERENCE NUMBERS.

☐ OVERALL CROSS-SECTION LINES AND KEYS.

☐ MATCH-UP LINE, OVERLAP LINE, AND REFERENCE IF FLOOR PLAN IS CONTINUED ON ANOTHER SHEET.

☐ SPACE FOR ITEMS N.I.C.____ NOTE ON N.I.C. ITEMS TO BE INSTALLED, CONNECTED BY CONTRACTOR____.

☐ NOTE SPACES WITH TWO-STORY OR HIGHER CEILINGS.

☐ REMODELING: EXISTING WORK TO BE REMOVED____ EXISTING WORK TO REMAIN____ EXISTING WORK TO BE REPAIRED OR ALTERED____.

☐ FRAMING, SHEATHING, AND SIDING NAILING SCHEDULE.

☐ GENERAL NOTES____ BUILDING CODE REFERENCES____.

DWELLING UNITS: PLANS AND INTERIOR ELEVATIONS

This includes items mainly found in custom or luxury housing. Otherwise the list is applicable to all types of housing.

ENTRY

☐ EXTERIOR LANDING, STOOP, OR PORCH: MATERIAL____ ELEVATION POINTS____ SLOPE____. (Nonskid finish surface if exposed to weather. Surface is 3″ below door threshold if exposed to weather.)

☐ RECESSED DOORMAT.

☐ WEATHER- AND SKID-RESISTIVE VESTIBULE FLOOR SURFACE.

☐ ARROW ON DRAWING TO INDICATE MAIN ENTRY____ SMALLER ARROW FOR SECONDARY ENTRY____.

☐ MAIN ENTRY AND SERVICE ENTRY ROOF OVERHANG OR CANOPY WEATHER PROTECTION.

☐ MAIL SLOT IN WALL OR DOOR, OR EXTERIOR-MOUNTED MAILBOX.

☐ GLAZING AT SIDE OF DOOR____ VIEWING LENS IN DOOR____.

☐ GUEST COAT CLOSET: HOOKS____ SHELF____ POLE____ WEATHER-RESISTIVE FLOOR-ING____.

☐ ENTRY VESTIBULE POWDER ROOM WITH LAVATORY AND TOILET. (Sound-isolation wall construction and door stripping.)

☐ SNOW COUNTRY: STORM VESTIBULE WITH WEATHERPROOF FLOOR____ SNOW GRATING AND DRAIN____ BENCH____ WET CLOTHING AND SKI EQUIPMENT STORAGE____.

☐ BEACH HOUSE: EXTERIOR SHOWER____ DRESSING ROOM AND LAUNDRY NEAR BEACH ENTRY____ BATHROOM WITH EXTERIOR ACCESS____ EXTERIOR DRINKING FOUNTAIN____.

MAIN LIVING AREAS: LIVING ROOM, FAMILY ROOM, DINING ROOM

☐ MASONRY OR PREFAB FIREPLACE____ FLUE SIZE____ FIREWOOD STORAGE____ EXTERIOR ASH DUMP CLEANOUT____ HEARTH SIZE AND MATERIAL____ HEIGHT OF RAISED HEARTH____ GAS FIRE STARTER PIPE WITH KEY VALVE AT HEARTH____ FIREPLACE DETAIL REFERENCE____.

☐ BUILT-IN SEATING.

☐ WET BAR____ SMALL UNIT REFRIGERATOR____ GLASSWARE CABINET____ BAR TOP MATERIAL____ CABINET WITH LOCK____.

☐ DINING BAR WITH PASS-THROUGH TO KITCHEN____ BAR TOP MATERIAL____ BUILT-IN WARMING PLATE____.

☐ HI-FI____ TV____ GAMES CABINET____ DARTBOARD WALL____.

☐ CHAIR RAIL (usually at formal dining room)____ PICTURE MOLDING OR RAIL____ WAINSCOT____ PANELING____.

☐ WALL-MOUNTED DISPLAY CABINETS____ SHELVING____ BUILT-IN TROPHY CASE____ HOBBY CASE____ GUN CASE____.

☐ INTERIOR PLANTERS: LINER MATERIAL____ DRAINS____ WATER SUPPLY____ DETAIL KEYS____.

☐ GUARDRAILS AT FULL-HEIGHT GLASS WALLS.

KITCHEN AND DINETTE

☐ BASE CABINETS AND COUNTER TOP____ COUNTER TOP AND SPLASH MATERIALS____.

☐ BROKEN-LINE INDICATION OF OVERHEAD CABINETS.

☐ CENTRAL WORK ISLAND.

☐ CUTTING BOARD. (Slide-out breadboard or chopping block counter top surface.)

☐ PANTRY CLOSET____ SPICE CABINET OR SHELF____ TRAY STORAGE BINS____ DRAWERS____ POT AND PAN RACK____.

☐ BROOM OR CLEANING CLOSET WITH SHELF.

☐ PLANNING CENTER: COOKBOOK SHELVING___ RECIPE FILE___.

☐ SERVING AND CLEANUP CART RECESS.

☐ EQUIPMENT: SINK AND VEGETABLE SINK___ BOILING WATER OUTLET___ HAND LOTION DISPENSER___ DISHWASHER WITH COUNTER TOP AIR GAP DEVICE___ GARBAGE DISPOSAL___ TRASH COMPACTOR___ RANGE OR COOKTOP___ AUXILIARY COOKTOP___ BUILT-IN FOOD WARMER___ INDOOR BBQ WITH EXHAUST___ BUILT-IN OVEN___ RANGE OR COOKTOP EXHAUST HOOD WITH LIGHT AND FAN___ SPACE FOR ELECTRONIC OVEN___ BUILT-IN, MULTIPURPOSE MIXING CENTER APPLIANCE___ SPACE FOR REFRIGERATOR/FREEZER WITH CABINET OR SOFFIT OVER___.

☐ GAS OUTLETS AND ACCESSIBLE SHUTOFFS FOR APPLIANCES: REFRIGERATOR___ OVEN___ RANGE___ BROILER OR INDOOR BBQ___.

☐ COLD WATER STUB AND VALVE (⅜″) FOR REFRIGERATOR AUTOMATIC ICE MAKER.

☐ VENTS AND DUCT TO OUTSIDE AIR FOR RANGE HOOD AND OVEN. (Concealed duct shown in broken line.)

☐ BULLETIN BOARD___ CHALKBOARD___.

☐ RECESS FOR KITCHEN CLOCK.

☐ PASS-THROUGH AND/OR DINING BAR AT KITCHEN WALL ADJACENT TO TERRACE OR PATIO.

☐ DOUBLE-ACTING GATE ("RESTAURANT DOOR") AT KITCHEN-DINING DOORWAY.

☐ INTERIOR ELEVATION ARROW SYMBOLS___ DETAIL KEYS___.

☐ INTERIOR ELEVATIONS: CABINET DOOR SWINGS___ SHELVING___ DRAWERS___ RACKS___ PULL-OUT EQUIPMENT___ CABINETS TO CEILING OR TO SOFFIT___ BASE, COVE, AND TOE SPACE HEIGHT AND MATERIAL___ SPLASH MATERIAL AND HEIGHT___ DOOR AND WINDOW SURROUNDS AND TRIM___ COUNTER TOP MATERIAL AND HEIGHT___ HEIGHT OF CABINETS ABOVE COUNTER___ HEIGHT OF VENT HOOD___ ELECTRIC OUTLETS AND SWITCHES___. (See Electrical Plan, Kitchen and Dinette, for other items.)

☐ DINETTE CHAIR RAIL___ WAINSCOT___.

UTILITY ROOM, MUD ROOM, LAUNDRY

☐ WASHER AND DRYER WITH CABINET AND LIGHT OVER.

☐ DRYER VENT TO OUTSIDE AIR.

☐ WATER SOFTENER.

☐ LAUNDRY SINK AND DRAINBOARD___ LAUNDRY STANDPIPE___.

☐ WATER-RESISTIVE FLOOR SURFACE AT LAUNDRY___ FLOOR DRAIN___.

☐ WATER HEATER WITH CAPACITY IN GALLONS___ RELIEF VALVE PIPE AND FLOOR DRAIN___ EXHAUST FLUE, MATERIAL AND SIZE___. (Water heater should be located to avoid proximity to parked cars, gasoline, or other volatile fumes.)

☐ GAS WATER HEATER CABINET COMBUSTION AIR VENTS AT FLOOR AND CEILING.

☐ GAS SUPPLY AND ACCESSIBLE SHUTOFF VALVES FOR GAS WATER HEATER AND CLOTHES DRYER.

☐ FREEZER OR AUXILIARY REFRIGERATOR SPACE: GAS STUB-OUT AND SHUTOFF.

☐ IRONING BOARD AND IRON STORAGE CABINET.

☐ CLOTHES SORTING BENCH.

☐ HANGING ROD, DRIP-DRY AREA, OR VENTILATED DRYING CLOSET.

☐ LINEN CLOSET. (Six shelves are typical.)

☐ BUILT-IN CLOTHES HAMPER. (Ventilated, sometimes supplied with small warm-air branch duct.)

☐ SEWING CENTER WORKBENCH AND CABINETS___ PEG- AND TACKBOARD___.

☐ COAT CLOSET OR CABINET NEAR SERVICE ENTRY DOOR.

☐ MOP AND SUPPLY STORAGE CABINET OR CLOSET.

☐ DOG OR CAT ENTRY DOOR IN EXTERIOR DOOR.

HALLWAYS

☐ HALL LINEN CLOSET (18″ deep; 6 shelves are typical).

☐ SKYLIGHTS OVER INTERIOR HALLWAYS AND STAIRWAYS.

☐ SCUTTLE OR FOLDING STAIR TO ATTIC.

STAIRWAYS

☐ STAIR RULES: MINIMUM WIDTH 30″ FOR BUILDING OCCUPANT LOAD FEWER THAN 10; 36″ FOR OCCUPANT LOAD OF 50 OR FEWER; 44″ FOR OCCUPANT LOAD OF MORE THAN 50.

☐ TREADS AND RISERS: RISERS 7½″ MAXIMUM; TREADS 10″ WIDE MINIMUM.

☐ LANDINGS: MINIMUM ONE EACH 12′ OF VERTICAL STAIR RUN___ MINIMUM LANDING

WIDTH, IN DIRECTION OF TRAVEL EQUAL TO STAIR WIDTH____.

☐ HANDRAILS: NOT REQUIRED FOR STAIRS WITH THREE OR FEWER RISERS____ AT LEAST ONE HANDRAIL FOR RESIDENTIAL STAIRS 44″ WIDE OR LESS____.

☐ HANDRAIL OR OTHER PROJECTION INTO STAIR AREA MAY REDUCE REQUIRED CLEARANCE WIDTH BY 3½″ MAXIMUM.

☐ HANDRAIL ENDS TERMINATE ON NEWELS OR RETURN TO WALL.

☐ HANDRAIL HEIGHT 30″ MINIMUM AND 34″ MAXIMUM ABOVE TREAD NOSINGS.

☐ HEAD ROOM: MINIMUM 6′6″ CLEARANCE BETWEEN NOSING AND SOFFIT.

☐ UNDER-STAIR STORAGE CABINETS.

☐ UNDER-STAIR SOFFIT MATERIAL.

☐ TREAD AND RISER MATERIALS____ SIZE OF NOSING PROJECTION____.

☐ BREAK LINE FOR PARTIAL PLAN OF STAIR ABOVE OR BELOW.

☐ NUMBER OF RISERS____ DIRECTION ARROW POINTING DOWN FROM FLOOR LEVEL____.

☐ STAIR AND LANDING DIMENSIONS____ FLOOR MATERIAL____ ELEVATION POINTS OF LANDINGS____.

BEDROOMS AND DRESSING ROOMS

☐ BUILT-IN HEADBOARD SHELF____ HEADBOARD CABINET____.

☐ BUILT-IN BED____ WATER BED____ CANOPY____ BED CEILING MIRROR____. (Ceiling mirror usually plastic tile or Mylar.)

☐ FULL-LENGTH MIRROR____ TWO- OR THREE-HINGE PANEL DRESSING MIRROR____.

☐ MAKEUP COUNTER OR VANITY INCLUDING: DRAWER____ CABINET____ LIGHTED MAKEUP MIRROR____.

☐ BUILT-IN VENTILATED HAMPER.

☐ LINEN CLOSET.

☐ LUGGAGE AND BLANKET STORAGE.

☐ WARDROBES WITH SHELVES AND RODS (two shelves if door is floor to ceiling height). (Minimum 2′2″ closet width is recommended; closet shelf 5′6″ high.)

☐ BUILT-IN VENTILATED DRESSER DRAWER OR BIN UNITS.

☐ FLOOR OR WALL SAFE. (Closet safes are well known to burglars; out-of-the-way location is recommended.)

☐ CEDAR PANELING AT WARDROBE.

☐ BOOK SHELVING____ STEREO AND TV CENTER____.

BATHROOMS

☐ FIXTURES: WATER CLOSETS____ LAVATORIES____ BIDET____ SHOWER____ TUB____ "JAPANESE" TUB____.

☐ TUB OR SHOWER ENCLOSURE: ROD AND DRAPE____ RIGID PLASTIC____ WIRE GLASS____. (Large, built-in combination "Japanese" bath/shower may not need enclosure.)

☐ TILE OR OTHER WATERPROOF WALL SURFACE: MATERIAL____ HEIGHT____.

☐ NOTE WATERPROOF CONSTRUCTION AT WATER-EXPOSED WALLS AND FLOORS.

☐ SHOWER SEAT____ ACCESSORIES SHELF____ SOAP AND SHAMPOO HOLDER NEAR SHOWER HEAD____.

☐ FLEXIBLE "TELEPHONE" SHOWER AND HOOK AT TUB____ TUB SHELF FOR BATH OILS AND POWDERS____.

☐ FORCED-AIR HEAT DUCT BRANCH TO UNDER-TUB SPACE.

☐ ACCESS PANEL FOR BATHTUB PLUMBING (14″ × 14″ is typical).

☐ BUILT-IN VENTILATED CLOTHES HAMPER.

☐ LAVATORY COUNTER, COUNTER TOP, AND SPLASH: MATERIALS____ HEIGHTS____.

☐ MEDICINE CABINET AND LIGHT.

☐ VANITY COUNTER WITH DRAWER AND CABINET.

☐ VANITY MAKEUP MIRROR.

☐ FULL-LENGTH MIRROR.

☐ OVERHEAD CABINET ABOVE WATER CLOSET.

☐ BATH SCALE SPACE OR BUILT-IN FLOOR SCALE.

☐ SKYLIGHT AT INTERIOR BATH.

☐ EXHAUST FAN VENT DUCT AT INTERIOR BATH.

☐ ACCESSORIES: ROBE HOOKS____ TOWEL BARS____ SOAP HOLDERS AT LAVS____ SOAP DISH AND GRAB BAR AT TUB AND SHOWER____ TOILET PAPER HOLDERS____ HAND LOTION DISPENSER____ TUMBLER HOLDER OR PAPER CUP DISPENSER____ TOOTHBRUSH HOLDER____.

☐ BUILT-IN PLANTER BOXES.

☐ BASE, FLOOR COVE, COUNTER TOE SPACE: MATERIALS____ HEIGHTS____.

☐ INTERIOR ELEVATION ARROW SYMBOLS____ DETAIL KEYS____.

FURNACE ROOM

- ☐ FURNACE: SIZE____ CLEARANCES FROM WALLS____ Btu RATING____ FLUE MATERIAL AND SIZE____.
- ☐ FURNACE CLOSET COMBUSTION AIR VENTS NEAR FLOOR AND CEILING (usually 1 sq. ft. for each 2000 Btu input rating).
- ☐ RETURN AIR DUCT.
- ☐ GAS OR FUEL SUPPLY AND SHUTOFF VALVE.
- ☐ INCOMBUSTIBLE PLATFORM FOR FORCED-AIR FURNACE ON WOOD FRAME FLOOR.

BASEMENT

See Basement Plan. Also see Utility Room, Furnace Room, Shop, and other related spaces in this section for items that might be included in a residential basement plan.

GARAGE, CARPORT, AND SHOP

- ☐ FLOOR SLAB. (See Concrete Slab on Grade.)
- ☐ FLOOR SLAB: SLOPE TO APRON OR TO DRAIN (2″ slope is typical)____ APRON GUTTER OR DRAIN IF DRIVEWAY SLOPES TOWARD GARAGE ENTRY____.
- ☐ POSTS: SIZE AND MATERIAL____ CONCRETE PEDESTAL FOR POSTS____.
- ☐ CORNER GUARDS, PROTECTIVE JACKETS AT POSTS, DOOR JAMBS, OR OTHER PARTS OF STRUCTURE SUBJECT TO DAMAGE BY VEHICLES.
- ☐ OVERHEAD DOOR: DOOR TYPE AND SIZE____ SIZE AND MATERIAL OF DOOR HEADER____ AUTOMATIC DOOR OPENER/CLOSER____.
- ☐ CEILING BRACES: DIAGONAL, FLAT TWO-BY-FOURS AT WOOD-FRAME OVERHEAD DOOR OPENING.
- ☐ GARAGE EXIT DOOR: EXTERIOR SWINGING, WITH OVERHANG AND SHELTERED ACCESS TO MAIN-BUILDING ENTRY.
- ☐ ATTACHED GARAGE: SOLID-CORE DOOR WITH AUTOMATIC CLOSER, BETWEEN GARAGE AND MAIN BUILDING.
- ☐ GARAGE WALLS: ONE-HOUR FIRE-RESISTIVE CONSTRUCTION AT WALLS COMMON TO MAIN BUILDING.
- ☐ SCREENED VENTS TO OUTSIDE AIR. (Typical is 60 sq. in. of vent per car. Vents within 6″ of floor.)
- ☐ STORAGE CABINETS: WALL- AND CEILING-MOUNTED UNITS OVER CAR HOOD SPACE.
- ☐ SHOP WORKBENCH AND STORAGE CABINET.
- ☐ UTILITY SINK OR LAUNDRY SINK.
- ☐ FREEZER SPACE.
- ☐ HOSE BIBB FOR CAR WASHING.
- ☐ GAS OUTLET: CAPPED, WITH VALVE AT WORKBENCH.

GARDEN SHED, GREENHOUSE

- ☐ FLOWER SINK.
- ☐ TOOL RACK____ POT SHELVES____ RACKS FOR FLATS____.
- ☐ WORKBENCH AND CABINET.
- ☐ GAS OUTLET: CAPPED, WITH VALVE AT WORKBENCH.
- ☐ HOSE BIBBS AND GARDEN HOSE STORAGE REELS.

EXTERIOR

See Site Work and Appurtenances.

- ☐ COURTYARD____ ATRIUM____ PAVING, SLOPE, AND DRAINS____.
- ☐ STORAGE CABINETS: RECREATIONAL EQUIPMENT____ GARDEN FURNITURE____ TOOLS____.
- ☐ GARDEN HOSE STORAGE REELS.
- ☐ TRASH CANS: BURIED RECEPTACLES____ RACK____ STORAGE CABINET____.
- ☐ BBQ GAS OUTLET.
- ☐ HOSE BIBBS/SILL COCKS____ FROSTPROOF BIBBS____.

ELECTRICAL PLAN: GENERAL REQUIREMENTS

A room-by-room survey of residential electrical components is provided under Electrical Plan: Residential Components.

- ☐ SERVICE ENTRY____ ELECTRIC METER____ SERVICE ENTRANCE SWITCH____. (Typical single-family residential service would be three-wire, 240 volts, single-phase.)
- ☐ CIRCUIT BREAKER BOX (30″ clear space in front of each box).
- ☐ BRANCH CIRCUIT PANEL BOXES.
- ☐ CIRCUIT PLANNING. (General rule allows one circuit for each 500 sq. ft. of floor space, one circuit for each major appliance, plus provision for future expansion.)
- ☐ CONVENIENCE OUTLETS. (Located along walls of habitable rooms at average intervals of 12′, minimum three per room. Typical height is 12″ centered above floor.)
- ☐ OUTLETS FOR CLEANING EQUIPMENT: IN CORRIDORS, HALLS, AND AT STAIRS.

☐ OUTLETS LOCATED ABOVE COUNTER TOPS OR AT OTHER NONSTANDARD HEIGHTS ARE NOTED IN INCHES ABOVE FLOOR___ OUT-LETS ARE NOT LOCATED BACK TO BACK IN SAME STUD SPACES WITHOUT ADDED BARRI-ERS TO NOISE TRANSMISSION BETWEEN ROOMS___.

☐ SWITCHES: TYPICALLY LOCATED AT EACH ROOM ACCESS OPPOSITE THE DOOR SWING, WITH THREE- AND FOUR-WAY SWITCHES AT STAIRS, HALLS, AND MULTIPLE-ENTRY ROOMS___ TYPICAL HEIGHT IS 48″ CENTERED ABOVE FLOOR___ NONTYPICAL HEIGHTS NOTED ON PLAN___.

☐ RESIDENTIAL LIGHTING: SWITCHED CEILING OR COVE LIGHTS USUALLY PROVIDED IN MA-JOR-USE AREAS, SUCH AS KITCHEN AND DIN-ING ROOMS. OTHERWISE, SWITCHES CON-TROL ONE-HALF OF ONE OR MORE DUPLEX OUTLETS NEAR FLOOR OR TABLE LAMP LOCA-TIONS.

☐ TIMER LIGHT SWITCH: TENANT GARAGE___ INTERIOR STAIR WELLS___ CORRIDORS___ YARD LIGHTS___ YARD SPRINKLERS___.

☐ BUILT-IN VACUUM SYSTEM___ MAIN SWITCH AT VACUUM MOTOR AND TANK___ VACUUM OUTLET SWITCHES___.

☐ DOOR-ACTIVATED SWITCHES OR PULL-CHAIN SWITCHES AT LARGE INTERIOR CLOSETS OR UTILITY AREAS.

☐ TELEPHONE PANEL BOARD___ TELEPHONE HOOKUPS___ JACKS FOR PORTABLE PHONES___.

☐ TV/FM CABLE OR ANTENNA LINE JACKS NEAR APPROPRIATE ELECTRIC CONVENIENCE OUT-LETS.

☐ INTERCOMS___ HEIGHT ABOVE FLOOR NOT-ED ON PLANS___.

☐ SMOKE AND HEAT DETECTORS AND ALARMS.

☐ INTRUDER DETECTORS AND ALARMS.

☐ 220/240 CIRCUITS AND OUTLETS FOR ROOM AIR-CONDITIONING UNITS.

☐ 220/240 CIRCUITS AND OUTLETS FOR ELEC-TRIC UNIT HEATERS.

ELECTRICAL PLAN: RESIDENTIAL COMPONENTS

This list includes electrical features found in luxury housing. Otherwise, the list is applicable to housing of all types.

ENTRY

☐ DOORBELL. (Double chimes at main entry, single chime at secondary entry. Chimes and transformer usually in hall or kitchen.)

☐ EXTERIOR INTERCOM.

☐ DOOR OR GATE ELECTRIC LOCK CONTROL WITH BUZZER.

☐ ADDRESS LIGHT.

☐ EXTERIOR LIGHTS: SOFFIT OR WALL-MOUNT-ED LIGHTS FOR FRONT, REAR, AND BASEMENT ENTRIES.

☐ EXTERIOR WEATHERPROOF CONVENIENCE OUTLETS NEAR FRONT AND REAR ENTRIES.

☐ ENTRY VESTIBULE CEILING LIGHT.

☐ SWITCHES: VESTIBULE LIGHT___ EXTERIOR LIGHTS___ YARD OR CARPORT LIGHTS___ ADJACENT LIVING ROOM, FAMILY ROOM, OR HALL LIGHTS___.

MAIN LIVING AREAS: LIVING ROOM, FAMILY ROOM, DINING ROOM

☐ TV/FM ANTENNA OR CABLE JACKS.

☐ TELEPHONE JACKS.

☐ DINING ROOM CEILING LIGHT CENTERED OVER DINING TABLE LOCATION. (Provision for con-cealed conduit in plank-and-beam ceiling.)

☐ LIGHT SWITCHES: ROOMS___ EXTERIOR BAL-CONY___ PATIO OR DECK LIGHTS___.

☐ DIMMER SWITCHES. (Dimmer switch may inter-fere with nearby stereo system.)

☐ CONVENIENCE OUTLETS: AT WET BAR___ AT DINING BAR___ NEAR DINING TABLE___ AT SERVING SHELF___.

KITCHEN AND DINETTE

☐ INTERCOM.

☐ KITCHEN CLOCK.

☐ OUTLETS: GAS OVEN LIGHT AND TIMER___ RANGE LIGHT AND TIMER___.

☐ WALL-MOUNTED TELEPHONE. (Light for note-taking.)

☐ DOOR CHIMES AND TRANSFORMER.

☐ HOOD, FAN, AND LIGHT OVER RANGE OR COOKTOP.

☐ TRASH COMPACTOR OUTLET. (Separate #10 copper wire circuit for each major appliance.)

☐ UNDER-COUNTER OUTLET FOR BOILING WA-TER APPLIANCE.

☐ HOOKUP FOR BUILT-IN, MULTIPURPOSE MIX-ING CENTER APPLIANCE.

☐ COUNTER SWITCH FOR GARBAGE DISPOSAL IF UNIT DOES NOT INCLUDE BUILT-IN SWITCH. (Under-counter outlet or junction box hookup.)

☐ COUNTER TOP CONVENIENCE OUTLETS, SPACED AT AVERAGE 12″ INTERVALS.

☐ 220/240 CIRCUITS____ OUTLET LOCATIONS SHOWN FOR ELECTRIC RANGE AND/OR OVEN.

☐ REFRIGERATOR AND FREEZER OUTLETS.

☐ CEILING LIGHTS: SINK____ FOOD PREPARATION AREAS____ DINETTE____.

☐ LIGHTS WITH INTEGRATED SWITCHES BENEATH OVERHEAD CABINETS.

☐ PANTRY LIGHT.

☐ LIGHT AT RECIPE AND PLANNING CENTER.

☐ WALL-MOUNTED SWITCHES FOR CEILING LIGHTS____ THREE- OR FOUR-WAY SWITCHES FOR LIGHT CONTROL OF OTHER AREAS____.

☐ EXTRA ELECTRICAL PROVISION WHERE NECESSARY: ROTISSERIE____ WARMING PLATE____ MEAT SLICER____ COFFEE GRINDER____ ELECTRONIC OVEN____ POPPER____ COUNTER TOP BROILER EXHAUST FAN____ WALL-MOUNTED CAN OPENER____ WRAP DISPENSER____ BUILT-IN TOASTER____ ICE MAKER____.

UTILITY ROOM, MUD ROOM, LAUNDRY

☐ CONVENIENCE OUTLETS FOR WASHER, DRYER (220/240 circuit and outlet for electric dryer).

☐ OUTLET FOR WATER SOFTENER.

☐ OUTLET FOR FREEZER OR AUXILIARY REFRIGERATOR.

☐ 220/240 OUTLET FOR ELECTRIC WATER HEATER.

☐ OUTLET FOR IRON.

☐ CEILING OR WALL EXHAUST FAN____ FAN SWITCH____.

☐ LIGHTS AT LAUNDRY AND CLOTHES SORTING AREAS.

☐ LIGHT AND OUTLETS AT SEWING CENTER.

☐ EXTERIOR LIGHT AT OUTSIDE ENTRY____ THREE-WAY SWITCH TO BREEZEWAY, GARAGE, OR CARPORT LIGHT____. (Lighted switches at rear entry door.)

HALLWAYS AND STAIRWAYS

☐ THERMOSTAT NEAR MAIN LIVING AREAS. (See Heating and Ventilating Plan.)

☐ OUTLETS FOR FLOOR CLEANING EQUIPMENT.

☐ SWITCHES FOR CENTRAL VACUUM SYSTEM.

☐ CEILING LIGHTS NEAR LINEN CLOSETS, OTHER HALL STORAGE.

☐ CEILING LIGHTS AND THREE- OR FOUR-WAY SWITCHES AT EACH END OF HALL OR STAIR RUN.

☐ HEAT AND SMOKE DETECTORS (near return air grille if using forced-air heating system).

BEDROOMS AND DRESSING ROOMS

☐ TELEPHONE JACKS.

☐ TV/FM ANTENNA OR CABLE JACKS.

☐ INTERCOM.

☐ BED HEADBOARD OUTLETS FOR CONVENIENCES: CLOCK RADIO____ ELECTRIC BLANKET____ STEREO SYSTEM____ READING LIGHTS____ MASSAGER____ TV____.

☐ OUTLETS ADEQUATE FOR ADDITIONAL APPLIANCES: SHOE BUFFER____ HAIR DRYER____ STEAM FACIAL APPLIANCE____ VAPORIZER____.

☐ YARD LIGHT MASTER CONTROL SWITCH.

☐ WARDROBE OR WALK-IN CLOSET LIGHTS. (Wall switch, pull-chain, or door-activated switch.)

☐ MAKEUP MIRROR AND DRESSING MIRROR LIGHTS.

BATHROOMS

☐ TELEPHONE JACK.

☐ UNIT HEATER SWITCH.

☐ EXHAUST FAN AND SWITCH.

☐ LIGHT AND OUTLET AT MEDICINE CABINET.

☐ OUTLETS ADEQUATE FOR ADDITIONAL APPLIANCES: ELECTRIC RAZOR____ HAIR STYLER____ HAIR DRYER____ ELECTRIC TOOTHBRUSH____ DENTURE CLEANER____ HYGIENE EQUIPMENT____ MASSAGER____ RADIO____ CLOCK____ WHIRLPOOL BATH____.

☐ MIRROR LIGHTS: FULL-LENGTH MIRROR____ MAKEUP MIRROR____.

☐ SWITCHES AND OUTLETS LOCATED OUT OF REACH OF TUB, SHOWER, AND LAVATORIES (as required by code or by normal safety considerations).

FURNACE ROOM

☐ FURNACE CONTROL SWITCH.

☐ CEILING OR WALL LIGHT TO ILLUMINATE FRONT OF FURNACE.

ATTIC

☐ CEILING LIGHT(S) WITH WALL SWITCH OR PULL CHAIN.

☐ CONVENIENCE OUTLET: ONE MINIMUM____ CIRCUIT AND JUNCTION BOX FOR FUTURE EXPANSION____.

BASEMENT

See Electrical Plan: General Requirements. See also Utility Room, Furnace Room, Shop, and other rooms that might be located in the basement.

☐ SUMP PUMP AND SWITCH AT FLOOD AREAS.

☐ SEWER PUMP, SWITCH, AND ALARM AT INSTALLATIONS BELOW PUBLIC SEWER INLET.

☐ TEMPERATURE AND HUMIDITY CONTROLS FOR RED AND WHITE WINE CELLARS.

☐ SAUNA HEATER.

☐ BASEMENT LIGHTING: ONE CEILING OUTLET MINIMUM FOR EACH 150 SQ. FT.

GARAGE, CARPORT, AND SHOP

☐ OUTLET AT FREEZER OR AUXILIARY REFRIGERATOR SPACE.

☐ INTERCOM.

☐ 220/240 SHOP TOOL OUTLETS.

☐ WORKBENCH OUTLETS____ STRIP OUTLETS____.

☐ WORKBENCH LIGHT AND SWITCH.

☐ CONVENIENCE OUTLETS, ADEQUATE FOR THREE TO FOUR APPLIANCES: VACUUM CLEANER____ COMPRESSOR____ ELECTRIC HEATER____ TROUBLE LIGHTS____ POWER TOOLS____.

☐ AUTOMATIC GARAGE DOOR OPENER.

☐ CEILING LIGHTS WITH OUTLETS____ THREE- OR FOUR-WAY LIGHTED SWITCHES FOR YARD LIGHTS, BUILDING ENTRY LIGHTS____.

☐ UTILITY LIGHT AND OUTLET FOR HIGH-CEILING UNEXCAVATED SPACES, WALK-IN STORAGE SPACES, AND CRAWL SPACES.

EXTERIOR

☐ WEATHERPROOF CONVENIENCE OUTLETS WITH GROUND FAULT DETECTORS FOR OUTDOOR EQUIPMENT: BBQ MOTOR____ POWERED GARDENING EQUIPMENT____ POOL EQUIPMENT____ WET/DRY VACUUM CLEANER____ INSECT TRAPS OR CONTROL LIGHTS____.

☐ WEATHERPROOF SPOT- OR FLOODLIGHTING: GARAGE ENTRANCE____ BUILDING ENTRANCES____ DECKS____ PATIO____ POOL____ SERVICE YARD____. (Switch may be controlled manually, by timers, or by electric eyes.)

☐ LIGHTS FOR MISCELLANEOUS AREAS: BALCONIES____ TOOL SHED____ GARDEN SHED____ GAZEBO____ TRASH AREA____ OTHER AUXILIARY STRUCTURES____.

☐ DECORATIVE YARD LIGHTING WITH TRANSFORMER____ TIMER____ CONTROL SWITCHES____.

☐ FOUNTAIN PUMP AND CONTROL SWITCH.

☐ AUTOMATIC YARD SPRINKLER SYSTEM____ TIMER____ CONTROL SWITCH____.

☐ SIDEWALK AND DRIVEWAY SNOW- AND ICE-MELTING EQUIPMENT.

☐ GUTTER DEICERS.

☐ OUTLET FOR ELECTRIC HEAT TAPE FOR WATER SUPPLY PIPE EXPOSED TO FREEZING TEMPERATURES.

☐ WATER WELL PUMP AND CONTROL SWITCH.

☐ JUNCTION BOX FOR AIR-CONDITIONING/HEAT PUMP EQUIPMENT.

HEATING AND VENTILATING PLAN

☐ DUCTWORK PLAN OR STEAM RISER DIAGRAM. (Provided by consultants for larger buildings. May be provided by contractor for smaller buildings.)

☐ RADIANT HEATING PIPING PLAN. (Provided by consultants or by heating contractor.)

☐ RETURN AIR GRILLE AND DUCTWORK.

☐ ELECTRIC OR GAS UNIT HEATERS: SIZES____ Btu RATINGS____.

☐ WARM-AIR DIFFUSERS____ STEAM OR HOT WATER REGISTERS____ BASEBOARD CONVECTORS (usually located at each large window of habitable rooms)____.

☐ THERMAL INSULATION: SPECIFIED OR NOTED ON DRAWINGS (usually shown in wall and other construction section details but not in plan).

☐ FUEL STORAGE AND DELIVERY FACILITIES.

☐ EXTERIOR HEAT EXCHANGER FOR AIR-CONDITIONING____ SUPPORT STRUCTURE OR SLAB____.

☐ ROOFTOP AIR INTAKES____ AIR-CONDITIONING EQUIPMENT AND SUPPORTING AND PROTECTIVE STRUCTURES____.

☐ FURNACE OR BOILER: LOCATION____ CLEARANCES TO WALLS FOR LEGAL AND MAINTENANCE REQUIREMENTS____ Btu INPUT-OUTPUT____.

☐ OUTSIDE AIR FOR FURNACE OR BOILER COMBUSTION.

☐ FLUES: SIZE____ MATERIAL____ DETAIL REFERENCE KEYS____.

☐ WARM-AIR HUMIDIFIER AND HUMIDISTAT____.

☐ DEHUMIDIFIER.

☐ AIR CLEANER.

☐ THERMOSTATS: 44″ FROM FLOOR.

☐ AIR-CONDITIONING UNITS AT EXTERIOR WALLS.

☐ HOT AIR AND WASTE AIR EXHAUSTS (located clear of usable yard areas).

☐ ROOF EXHAUSTS, VENTS, INTAKES____ LOCATED TO AVOID SNOW CLOGGING____ LOCATED TO PREVENT EXHAUSTS FROM ENTERING FRESH AIR INTAKES____.

☐ INTERIOR DIFFUSERS AND RETURN AIR VENTS LOCATED TO AVOID SHORT CIRCUITS IN AIR FLOW.

☐ EXTERIOR AIR-CONDITIONING EQUIPMENT LOCATED TO AVOID NOISE PROBLEMS FOR NEIGHBORS OR OCCUPANTS____ PROTECTED FROM WEATHER, ANIMALS, AND VANDALISM____.

PLUMBING PLAN

Some building code jurisdictions require a schematic diagram indicating pipe sizes to ensure compliance with minimal code requirements. In some locales, detailed plans of all piping are shown overlaid on floor plans. In others, separate plumbing plans are not provided; requirements are stated in plumbing note or in specifications.
See Plumbing Fixture Schedule.

TYPICAL WATER SUPPLY PIPE

Gas piping sizes are determined by local code and specific supply size requirements of appliances and fixtures.

☐ METER TO HOUSE: 1½″.

☐ HOUSE MAIN: 1″ TO 1½″.

☐ HOT AND COLD BRANCH MAINS: ¾″ OR 1″.

☐ HOSE BIBB SUPPLY: ½″ OR ¾″.

☐ FIXTURE BRANCHES TO SHOWER, KITCHEN SINK, DISHWASHER, BATHTUB, LAUNDRY: ½″.

☐ FIXTURE BRANCH TO LAVATORY: ⅜″ OR ½″.

☐ FIXTURE BRANCH TO WATER CLOSET: ⅜″.

WATER SUPPLY DETAILS

☐ WIDE CURVE SWEEPS RATHER THAN RIGHT-ANGLE TURNS RECOMMENDED FOR BRANCH MAINS, TO REDUCE TURBULENCE NOISE.

☐ SOLID SUPPORT BLOCKING AND CONNECTION AT CHANGES IN PIPE DIRECTION.

☐ NOISE-REDUCTION GASKETS AT CONTACT OF

PIPE WITH BUILDING FRAMING MEMBERS.

☐ MINIMUM 8″ DISTANCE BETWEEN HOT AND COLD WATER PIPES.

☐ FROSTPROOF VALVES WHERE EXPOSED TO FREEZING WEATHER.

☐ SAFETY TEMPERATURE CONTROL VALVE AT SHOWER.

☐ SHUTOFF VALVE AT EACH FIXTURE BRANCH____ GATE VALVE AT HOUSE MAIN____.

☐ AIR CHAMBERS AT FIXTURES: 12″ LONG (pipe one size larger than supply pipe).

☐ HOT WATER RETURN LOOP TO WATER HEATER FOR CONTINUOUS HOT WATER SUPPLY AT FIXTURES.

☐ WATER SUPPLY SLOPE TO DRAIN AT MAIN SHUTOFF GATE VALVE.

☐ WATER SUPPLY PIPE INSULATION.

☐ AIR GAP AT DISHWASHER.

☐ ELECTROLYTIC FITTINGS JOIN PIPES OF DISSIMILAR METALS.

TYPICAL SOIL AND WASTE PIPE

☐ WATER CLOSET: 3″ OR 4″.

☐ SHOWER, BATHTUB, KITCHEN SINK, GARBAGE DISPOSAL, LAUNDRY: 2″.

☐ LAVATORY: 1½″.

☐ FLOOR DRAINS: 3″ to 4″.

☐ HOUSE SEWER MAIN: 3″ TO 4″.

☐ SOIL AND WASTE CLEANOUTS: BOTTOM OF EACH STACK____ AT EACH CHANGE IN DIRECTION OVER 22½°____ EACH 50′ IN STRAIGHT RUNS____. (Horizontal drainpipe slopes ¼″ per ft.)

STACKS AND VENTS

☐ MAIN SOIL STACK, CONTINUOUS FROM WATER CLOSET OR FROM COMBINATION OF OTHER FIXTURES: 3″.

☐ SINGLE VENTS AND BRANCH VENTS FROM LAVATORIES, BATHTUB, SHOWER, KITCHEN SINK, LAUNDRY: 1½″.

EXTERIOR ELEVATIONS

SUBGRADE TO FLOOR LINE

- ☐ EXISTING AND NEW FINISH GRADE LINE AND ELEVATIONS (rough finish grade and topsoil depth sometimes shown)____ FILL AND ENGINEERED FILL____.
- ☐ ELEVATION POINTS AT BOTTOMS OF FOOTINGS.
- ☐ FOOTING, FOUNDATION WALL, AND BASEMENT WALL LINES BELOW GRADE: DASHED.
- ☐ EXISTING ROCK OUTCROPS TO BE REMOVED.
- ☐ FIREPLACE/CHIMNEY FOOTINGS.
- ☐ BUILDING SLAB FOOTING LINES____ SLAB FLOOR LINE____.
- ☐ DRAINPIPE AND GRAVEL BED AT FOOTINGS.
- ☐ BASEMENT WALL OPENINGS AND AREAWAYS____ AREAWAY GRILLE COVER OR GUARDRAIL____.
- ☐ FOUNDATION CRAWL SPACE VENTS AND ACCESS PANELS.
- ☐ LINE OF SLAB AT GARAGE OR CARPORT.
- ☐ SPLASH BLOCKS____ GRADE GUTTERS____ CURBS____.
- ☐ EXTERIOR WALKS____ LANDINGS____ STEPS____ PAVING WITH ELEVATIONS AND SLOPES____.
- ☐ FOOTINGS AND ROCK SUB-BASE FOR EXTERIOR SLABS____ LANDINGS____ STEPS____.
- ☐ SOIL AND CRUSHED-ROCK INDICATIONS.
- ☐ DIMENSIONS: FOOTING THICKNESSES____ FOUNDATION WALL HEIGHT____ DEPTHS OF AREAWAYS____ PAVING ELEVATIONS____ FLOOR ELEVATIONS____ BASEMENT FLOOR TO GROUND FLOOR____.
- ☐ CLEARANCES: CEMENT STUCCO OR MASONRY VENEER 8″ ABOVE FINISH GRADE____ STUCCO 12″ AWAY FROM SHRUBBERY____ UNTREATED WOOD SIDING MINIMUM 8″ FROM FINISH GRADE (unless separated from grade by minimum of 3″ of concrete)____ FLOOR JOISTS 18″ DISTANCE TO GRADE____ WOOD GIRDERS MINIMUM 12″ TO GRADE____.

FLOOR TO CEILING OR ROOF LINE

- ☐ BUILDING WALL OUTLINE____ OFFSETS____ CANOPIES____ BALCONIES____.
- ☐ DIMENSIONS: FLOOR TO CEILING HEIGHTS____ WINDOW, DOOR, OPENING HEAD HEIGHTS FROM FLOOR LINE____.
- ☐ NOTE WHETHER VERTICAL DIMENSIONS ARE TO FINISH SURFACES, SUBFLOORING, OR STRUCTURE____ WHETHER OPENINGS ARE ROUGH OR FINISH____.
- ☐ ADJACENT CURBS____ FENCING____ WALLS____ PLANTERS____ RETAINING WALLS____ PARTY WALLS____ PAVING____.
- ☐ WALL MATERIAL INDICATIONS, TEXTURES, PATTERNS, AND NOTES: BRICK OR BLOCK MASONRY COURSES AND HEADER COURSES____ STONE____ CONCRETE____ CEMENT PLASTER____ WOOD SIDING____.
- ☐ BATTENS: SIZES____ MATERIALS____ SPACING____.
- ☐ CONSTRUCTION JOINTS: EXPANSION AND CONTRACTION JOINTS____ CONTROL JOINTS IN MASONRY SPANDRELS OVER OPENINGS____ PLYWOOD SIDING MOVEMENT JOINTS (¼″ between panels)____.
- ☐ EXTERIOR STEPS____ STOOP, LANDING, AND HANDRAIL____ STOOP OR LANDING SLOPE AND ELEVATIONS (top of exterior landing 3″ below door threshold)____.
- ☐ DECK, PORCH, BALCONY LINES____ SLOPE AND ELEVATION POINTS____ RAILINGS WITH NEWELS OR HORIZONTAL MEMBERS SPACED AT 9″ MAXIMUM____.
- ☐ EXTERIOR-MOUNTED AIR-CONDITIONING AND VENTILATING EQUIPMENT.
- ☐ POSTS, COLUMNS: MATERIALS____ SIZES____.
- ☐ BUMPERS, WHEEL GUARDS, CORNER GUARDS: MATERIALS____ SIZES____ DETAIL KEYS____.
- ☐ EXTERIOR RAISED DECK FRAMING____ POSTS____ PIERS____ CROSS-BRACING____ STRUCTURAL CONNECTIONS____.
- ☐ TRIM: MATERIAL____ SIZE____ DETAIL KEYS____.
- ☐ SKIRT BOARDS: MATERIAL____ SIZE____ DETAIL KEYS____.
- ☐ WATER TABLE: MATERIAL____ SIZE____ DETAIL KEYS____.
- ☐ LEDGERS: MATERIAL____ SIZE____ DETAIL KEYS____.
- ☐ BARGEBOARDS: MATERIAL____ SIZE____ DETAIL KEYS____.
- ☐ DOWNSPOUTS____ LEADERS____ LEADER

BOOTS___.

□ HOSE BIBBS___ HYDRANTS AT WALL___ STANDPIPES___ SIAMESE FIRE DEPARTMENT CONNECTIONS AND IDENTIFYING SIGN___.

□ DOORS AND WINDOWS: SILLS___ TRIM___ MUNTINS___ MULLIONS___.

□ DOOR AND WINDOW SYMBOLS FOR SCHEDULE OF DOOR AND WINDOW TYPES AND SIZES. (Types and sizes may be noted on elevations of smaller projects without separate schedules.)

□ DOORS AND WINDOWS: DIRECTION OF SWING___ DIRECTION OF SLIDING UNITS___ SWING OF HOPPER UNITS___.

□ SHUTTERS.

□ FIXED GLASS AT NONOPERABLE WINDOWS___ JALOUSIES___.

□ OBSCURE GLASS AT BATHROOMS___ WIRE GLASS___ PLASTIC___.

□ LOUVERS/VENTS FOR LAUNDRY DRYER, RANGE HOOD, AND INTERIOR FORCED-AIR VENTILATED BATHS___ THRU-WALL AIR INTAKES___ BIRD OR INSECT SCREENS___.

□ BUILT-IN EXTERIOR DECK STORAGE___ HOSE REELS___ PLANT SHELVES___ TOOL RACKS___ FIREWOOD STORAGE___ DETAIL KEYS___.

□ EXTERIOR LIGHTS___ WEATHERPROOF CONVENIENCE OUTLETS___.

□ DOORBELL BUTTON___ INTERCOM___.

□ NAME AND ADDRESS PLAQUE.

□ MAIL SLOT OR WALL- OR POST-MOUNTED MAILBOX.

□ EXTERIOR-MOUNTED ALARMS.

□ TRASH CAN ENCLOSURE OR BURIED TRASH CONTAINER___ DETAIL KEYS___.

□ EXTERIOR GAS AND ELECTRIC METER ENCLOSURES___ ELECTRIC SWITCH BOX ENCLOSURE___.

□ DETAIL KEYS: POST CONNECTORS AND ANCHORS___ THRU-WALL OPENINGS___ SPECIAL SILL, JAMB, AND HEAD CONDITIONS___ ATTACHED PLANTERS___ CONSTRUCTED VENEERS___.

□ FLASHING AT CHANGES IN BUILDING MATERIAL.

□ DIAGONAL CROSS-BRACING IN WOOD FRAME WALLS. (Shown as dash and dot lines at 45° typical at each exterior corner and each 25 linear feet.)

□ PARTIAL DRAWING AND/OR NOTES: WALL SHEATHING___ MOISTURE BARRIER___ SIDING AND NAILING PATTERNS___. (May be covered in wall section details.)

□ DRIPS AT UNDERSIDE EDGE OF CANTILEVERED BALCONIES AND PROJECTING SILLS.

CEILING LINE TO ROOF

□ FASCIA: MATERIAL___ SIZE___ NOTE___ DETAIL KEYS___.

□ BARGEBOARD AT GABLES (also called "verge boards" or "rake boards")___ DETAIL KEYS___.

□ LOOKOUTS___ OUTRIGGERS___ BRACKETS___.

□ EXTERIOR SOFFIT LINE.

□ SOFFIT MATERIAL___ CONSTRUCTION AND CONTROL JOINTS___ DETAIL KEYS___. (Cement plaster soffits and control joints at 12' spacing and at building and fascia intersections.)

□ EAVE SOFFIT VENTS___ RIDGE VENT___ ATTIC VENTS AND LOUVERS___ LOUVER SIZES AND MATERIALS___.

□ ROOF GRAVEL STOP: GAUGE___ MATERIAL___ DETAIL KEY___.

□ CAP FLASHING: GAUGE___ MATERIAL___ DETAIL KEY___.

□ GUTTERS, BOX GUTTER, RAIN DEFLECTORS, DOWNSPOUTS, LEADER STRAPS, SCUPPERS: MATERIALS___ METAL GAUGES___ SIZES___.

□ PARAPET CAP MATERIAL (parapet cap slope inward toward roof)___ DETAIL KEYS___.

□ RAILINGS AT ROOF DECK___ FIRE ESCAPE OR EXTERIOR ROOF ACCESS LADDER AND RAILINGS___ DETAIL KEYS___.

□ FINISH ROOFING MATERIAL.

□ ROOF SLOPES.

□ SKYLIGHTS, MONITORS___ ROOF-MOUNTED EQUIPMENT___.

□ CHIMNEY AND CHIMNEY FLASHING SADDLE___ CRICKET___ COPING___ CAP___ FLUE___ SPARK ARRESTOR. (Fireplace chimneys, flues at 2' minimum above roof or above any roof projection within 10'.)

□ ANCHORS FOR TV, FM, OR SHORT-WAVE ANTENNAS___ LIGHTNING RODS___ WEATHER VANES___ OVERHEAD CABLE ENTRY HEADS___.

GENERAL REFERENCE DATA

□ SHEET TITLE AND SCALE.

□ TITLES OR SYMBOL KEYS TO IDENTIFY EACH ELEVATION VIEW.

□ EXISTING UNDERGROUND STRUCTURES: **253**

WHICH TO RETAIN____ WHICH TO RE-
MOVE____.

☐ HOLES, TRENCHES, EXCAVATIONS TO BE
FILLED.

☐ PROPERTY LINES____ SETBACK LINES____.

☐ OUTLINE OF ADJACENT STRUCTURES.

☐ OUTLINE OF FUTURE BUILDING ADDITIONS.

☐ BUILDING CROSS SECTION OR VERTICAL
WALL SECTION DETAIL KEYS.

WALL AND BUILDING SECTIONS

Residential and small-building detailing often combine foundation, wall framing elevation view, window sill and head sections, roof overhang and soffit,
and roof construction in a single partial building cross
section. Here's the typical ground-to-roof sequence for
such a drawing:

☐ EXISTING AND NEW FINISH GRADE LINES
WITH ELEVATIONS: TOPSOIL TO BE ADD-
ED____ SOIL INDICATIONS: UNDISTURBED
SOIL AND FILL____.

☐ FOOTING AND FOUNDATION WALL: MATERI-
ALS INDICATIONS____ REINFORCING____
THICKNESS AND HEIGHT DIMENSIONS____.

☐ FOOTING DRAIN TILE WITH GRAVEL BED AND
COVER-OVER TILE____.

☐ CRAWL SPACE____ CONCRETE RODENT BARRI-
ER____ VAPOR BARRIER WITH SAND COV-
ER____.

☐ SECTION AT BASEMENT AREAWAY: AREA-
WAY WALL AND FOOTING DIMENSIONS____
DRAIN____ WINDOW OR HATCH SECTION____.

☐ BASEMENT WALL MEMBRANE WATERPROOF-
ING____ DETAIL FOR PIPE AND SLEEVES THAT
PENETRATE WATERPROOFING____.

☐ BASEMENT OR FOUNDATION WALL GIRDER
RECESS____ METAL GIRDER BOX WHERE
WOOD FRAMING ENTERS WALL BELOW
GRADE LINE____.

☐ GIRDER, BEAM, AND JOIST AIR SPACE WHERE
WOOD MEMBERS ENTER OR CONNECT CON-
CRETE OR MASONRY FOUNDATION WALL OR
BASEMENT WALL.

☐ ADJACENT STOOPS, LANDINGS, OR OTHER
CONCRETE SLABS____ SLAB FOOTINGS OR
FROST CURBS____ TAMPED FILL, SAND, OR
GRAVEL FILL____ VAPOR BARRIER____ REIN-
FORCING____ SLAB THICKNESS____.

☐ DOWELS CONNECTING SLABS TO FOOTING OR
FOUNDATION WALL.

☐ TERMITE SHIELDS.

☐ FOUNDATION WALL TOP PLATE/MUD SILL:
TYPE AND SIZE____ .

☐ LEVELING GROUT FOR MUD SILL.

☐ MUD SILL ANCHOR BOLTS____ SIZE____
DEPTH____ SPACING____.

☐ MASONRY WALL: TYPE____ THICKNESS____

MATERIAL INDICATION___ AIR SPACE AND WEEP HOLES IF CAVITY WALL TYPE___ COURSES AND HEADER COURSES___ SILL FLASHING___ ANCHORS AND REINFORCING___ GROUTING___ PARGING___.

☐ MASONRY VENEER ON WOOD FRAME___ FOUNDATION WALL LIP SUPPORT FOR MASONRY___ BASE FLASHING___ ANCHORS TO WALL___ MATERIAL INDICATION___ COURSES AND CAP COURSE___ THICKNESS___ AIR SPACE OF 1″ BETWEEN MASONRY AND SHEATHING___ ANCHORS TO WALL___ HEAVYWEIGHT MOISTURE BARRIER___ SHEATHING___.

☐ FOUNDATION GIRDER SUPPORT PIERS: SIZES___ MATERIAL___ CAPS___ POSTS___.

☐ BASEMENT GIRDER SUPPORT POST OR COLUMN___ PIER FOUNDATION___ NONSHRINK GROUT BASE___ BASE PLATE AND FLOOR CONNECTORS FOR STEEL COLUMN___ PIER SUPPORT___ NONCORROSIVE METAL MOISTURE BARRIER WITH WOOD CAP FOR WOOD COLUMN___.

☐ GIRDERS, BEAMS, JOISTS, LEDGERS: MATERIALS___ WOOD GRADES___ SIZES___ TYPES AND LOCATION OF CONNECTORS___.

☐ JOIST SOLID BLOCKING OR CROSS-BRIDGING. (Note bottom nailing of bridging after subfloor is completed. Blocking required for each 8′ of joist span.)

☐ SUBFLOORING.

☐ FLOOR WATERPROOFING AT WET ROOMS.

☐ WOOD FRAME: BOTTOM PLATE, STUDS, AND DOUBLE TOP PLATE. (Double top plates overlap at 4′ minimum; no overlaps over openings.)

☐ LET-IN DIAGONAL BRACING. (Typical 45° bracing at corners, and at every 25′ of structure, especially near wall openings.)

☐ WALL SHEATHING. (Let-in diagonal bracing not normally required if wall is sheathed with ½″ plywood.)

☐ STUD SOLID BLOCKING. (As fire blocking where required by code or as nailing surface for horizontal plywood sheathing.)

☐ BUILDING FELT. (Typically 15# asphalt felt with 6″ overlaps.)

☐ EXTERIOR CEMENT PLASTER: LATH___ GROUNDS___ SCREEDS___.

☐ WINDOW SILL___ HEAD FRAMING___ TRIM___ STOOL___ APRON___.

☐ WALL AND CEILING VAPOR BARRIER___ RIGID OR BATT THERMAL INSULATION___.

☐ INTERIOR WALL FRAMING___ FINISH___ SEPARATED STUD SOUND WALL___ SOUND INSULATION___ WATERPROOFING AT WET ROOMS___.

☐ GIRT BOARD___ EXTERIOR BATTENS___ TRIM___.

☐ CEILING JOISTS AND BLOCKING___ FINISH CEILING MATERIAL___.

☐ ROOF JOISTS OR BEAMS___ RAFTERS___ PURLINS___ CROSSTIES___.

☐ OVERHANG___ FASCIA___ SOFFIT MATERIAL___ SOFFIT VENTS___ BLOCKING BETWEEN CEILING OR ROOF JOISTS AT TOP PLATE___ SOFFIT TRIM___.

☐ MASONRY WALL TOP PLATE OR LEDGER___ PLATE ANCHORS___ JOIST AND BEAM CONNECTORS___.

☐ ROOF DECKING OR SHEATHING___ FINISH ROOFING MATERIALS___ WATERPROOFING___.

☐ EDGE FLASHING___ GUTTER AND GUTTER CONNECTORS___ GRAVEL STOP___ CAP FLASHING___. (All flashing; material___ gauge___ if not covered in specifications.)

☐ ROOF RIDGE FLASHING___ HIP AND VALLEY FLASHING___.

☐ THRU-ROOF VENT PIPE AND FLUE FLASHING.

☐ PARAPET WATERPROOFING___ FLASHING REGLET___ FLASHING AND COUNTERFLASHING___.

☐ PARAPET CAP___ WASH TOWARD ROOF___ PARAPET EXPANSION AND CONSTRUCTION JOINTS___.

☐ NAILING SCHEDULE (may be with floor plan).

SCHEDULES

DOOR SCHEDULE

The simplest buildings require no schedule; data are noted at doors on the plan.

☐ DOOR TYPES DRAWN IN ELEVATION. (A scale of ½" is typical.)

☐ SYMBOL AND IDENTIFYING NUMBER AT EACH DOOR.

☐ DOOR WIDTHS____ HEIGHTS____ THICK-NESSES____.

☐ DOOR TYPES____ MATERIALS____ FINISH-ES____. (Number of doors of each type and size is sometimes noted.)

☐ OPERATING TYPE: SLIDING____ SINGLE-ACT-ING____ DOUBLE-ACTING____ DUTCH____ PIV-OT____ TWO-HINGE____ THREE-HINGE____.

☐ KICKPLATES.

☐ FIRE RATING IF REQUIRED.

☐ LOUVERS____ UNDERCUTS FOR VENTILA-TION____.

☐ SCREENS.

☐ MEETING STILES.

☐ DOOR GLAZING____ TRANSOMS____ BOR-ROWED LIGHTS____.

☐ DETAIL KEY REFERENCE SYMBOLS: SILLS____ JAMBS____ HEADERS____.

☐ MANUFACTURERS____ CATALOG NUM-BERS____. (If not covered in specifications.)

A common form of schedule for simpler buildings has no drawings and follows this sequence:

☐ DOOR SYMBOL AND IDENTIFYING NUM-BER____ WIDTH AND HEIGHT____ THICK-NESS____ TYPE____ MANUFACTURER AND MANUFACTURER'S NUMBER____ FINISH____ MISCELLANEOUS REMARKS____.

WINDOW SCHEDULE

As with the door schedule, a separate window sched-ule may not be required on the simplest buildings.

☐ WINDOW TYPES DRAWN IN ELEVATION. (A scale of ½" is typical.)

☐ WINDOW SYMBOL AND IDENTIFYING NUMBER AT EACH DRAWING.

☐ WINDOW SIZE____ WIDTH____ HEIGHT____.

☐ WINDOW TYPE____ DIRECTION OF MOVEMENT OF OPERABLE SASH AS SEEN FROM EXTERI-OR____. (Number of windows of each type and size is sometimes noted.)

☐ GLASS THICKNESS AND TYPE.

☐ SCREENS.

☐ NOTE: FIXED____ OBSCURE____ WIRE____ TEM-PERED____ DOUBLE GLAZING____ TINTED____.

☐ DETAIL KEY REFERENCE SYMBOLS: SILLS____ JAMBS____ HEADERS____.

☐ MANUFACTURERS____ CATALOG NUM-BERS____. (If not covered in specifications.)

A common form of schedule for simpler buildings has no elevation drawings of windows and follows this sequence:

☐ WINDOW SYMBOL AND IDENTIFYING NUM-BER____ WIDTH AND HEIGHT (note if dimen-sions are actual window frame size or rough open-ing size)____ WINDOW TYPE____ GLASS THICKNESS AND TYPE____ MATERIAL AND FINISH OF FRAME____ MANUFACTURER AND MANUFACTURER'S NUMBER____ MISCELLA-NEOUS REMARKS____.

FINISH SCHEDULE

A simple building may require no separate finish schedule; notes on the plan plus on-site instruction by the architect or owner are often adequate. Where the schedule is used, the following sequence is typical:

☐ ROOM NAME AND/OR IDENTIFYING NUMBER.

☐ FLOOR: MATERIAL____ FINISH____.

☐ BASE: HEIGHT____ MATERIAL____ FINISH____.

☐ WALLS: MATERIALS____ FINISHES____. (Walls may be identified by compass direction if finishes vary wall by wall. Note waterproofing and water-proof-membrane wall construction.)

☐ WAINSCOT: HEIGHT____ MATERIAL____ FIN-ISH____.

☐ CEILING: MATERIAL____ FINISH____.

☐ SOFFITS: MATERIAL____ FINISH____.

☐ CABINETS: MATERIAL SPECIES AND GRADE____ FINISH____.

☐ SHELVING: MATERIAL____ FINISH____.

☐ DOORS: MATERIAL____ FINISH____. (If not cov-ered in door schedule.)

☐ TRIM AND MILLWORK: MATERIAL SPECIES AND GRADE____ FINISH____.

☐ MISCELLANEOUS REMARKS OR NOTES.

☐ COLORS: STAIN AND PAINT____ MANUFAC-TURER AND TRADE NAMES OR NUMBERS (if not covered in specifications; may be left for later decision with provision for paint allowance by bidder)____.

☐ FINISHES: EXTERIOR WALLS____ SILLS____ TRIM____ POSTS____ FENCES____ GUTTERS AND LEADERS____ FLASHING AND VENTS____ DECKING____ SOFFITS____ FASCIAS____ RAILINGS____. (Included in finish schedule if not covered in specifications.)

ELECTRIC FIXTURE SCHEDULE

These data are usually provided in the specifications or in general note form on the electrical plan. Typical sequence:

☐ FIXTURE SYMBOL AND/OR NAME. (Quantity of each type is sometimes noted.)

☐ FIXTURE TYPE (If not named above).

☐ WATTS AND RATED VOLTAGE.

☐ MANUFACTURER AND TRADE NAME OR MANUFACTURER'S NUMBER.

☐ MISCELLANEOUS REMARKS OR NOTES.

PLUMBING FIXTURE SCHEDULE

The data are usually provided in the specifications or in general note form on the plumbing plan. Typical sequence:

☐ SYMBOL. (Symbols with letters or numbers may be used on plumbing plan to identify fixtures.)

☐ TYPE OR NAME OF FIXTURE. (Quantity of each fixture is sometimes noted.)

☐ SIZE.

☐ MANUFACTURER AND MANUFACTURER'S TRADE NAME OR NUMBER.

☐ WATER SUPPLY, DRAIN, AND VENT PIPE SIZES.

☐ MATERIAL____ FINISH____ COLOR____.

☐ MISCELLANEOUS REMARKS OR NOTES.

RESOURCES

Important Notice on
Update Information on Systems Graphics

New developments in systems graphics require a continuous update on sources of information and products. Guidelines is offering an annual supplement on the latest sources to keep readers up to date. To receive the current supplement, please send $4.00 per copy to *The Guidelines Reprographic and Computer Resources Guide*, Guidelines, Box 456, Orinda, CA 94563.

INDEX

About the Author

FRED A. STITT is an architect, writer, editor, publisher, and inventor with extensive experience in design and problem solving. Drawing on his practical experience in architectural offices and research at the College of Environmental Design at the University of California at Berkeley, he published more than a dozen practical booklets under the name of Guidelines Publications. Those first booklets broke new ground in architectural literature by filling the gap between academic knowledge of design and the actual needs of running a design practice.

Since 1969 Fred and his partner Marjorie Stitt have produced Guidelines Publications on architectural planning, creative problem solving, materials failures, A/E computerization, and many other subjects for architects and engineers. Fred is the author of the widely acclaimed *Systems Drafting* (McGraw-Hill, 1980), and currently has fifteen books in preparation. Since 1972 he has written *The Guidelines Letter*, the first successful newsletter on design firm management. He also writes articles for publications such as *Repro Graphics World* and *Architectural Record*. He lectures widely in the United States and abroad on such topics as the future of architecture, the origins of design consciousness, and breakthroughs in CADD and systems graphics both for universities and professional societies and for in-house seminars held by A/E firms.